Adam's Navel

ALSO BY MICHAEL SIMS

Darwin's Orchestra:
An Almanac of Nature in History and the Arts

Adam's Navel

A NATURAL AND CULTURAL HISTORY OF
THE HUMAN FORM

Michael Sims

VIKING

VIKING
Published by the Penguin Group
Penguin Group (USA) Inc., 375 Hudson Street, New York, New York 10014, U.S.A.
Penguin Books Ltd, 80 Strand, London WC2R 0RL, England
Penguin Books Australia Ltd, 250 Camberwell Road, Camberwell, Victoria 3124,
Australia
Penguin Books Canada Ltd, 10 Alcorn Avenue, Toronto, Ontario, Canada M4V 3B2
Penguin Books India (P) Ltd, 11 Community Centre, Panchsheel Park,
New Delhi–110 017, India
Penguin Books (N.Z.) Ltd, Cnr Rosedale and Airborne Roads, Albany, Auckland,
New Zealand
Penguin Books (South Africa) (Pty) Ltd, 24 Sturdee Avenue, Rosebank,
Johannesburg 2196, South Africa

Penguin Books Ltd, Registered Offices: 80 Strand, London WC2R 0RL, England

First published in 2003 by Viking Penguin, a member of Penguin Group (USA) Inc.

10   9   8   7   6   5   4   3   2   1

Copyright © Michael Sims, 2003
All rights reserved

Illustrations by Juliet Borda

CIP data is available.

ISBN 0-670-03224-7

This book is printed on acid-free paper. ∞

Printed in the United States of America

For
Heide Lange:
friend, agent, optimist

Is the dividing line between nature and culture the imagination?

GUY DAVENPORT

Although the intellectual powers and social habits of man are of paramount importance to him, we must not underrate the importance of his bodily structure. . . .

CHARLES DARWIN

Our being seems to lie not in cells and muscles but in the traces that our thoughts and actions inscribe on the air.

JOHN UPDIKE

The Ethiopians say that their gods are snub-nosed and black, the Thracians that theirs have light blue eyes and red hair.

XENOPHANES

CONTENTS

The Form Complete

Of physiology from top to toe I sing,
Not physiognomy alone nor brain alone is worthy for the Muse,
I say the Form complete is worthier far,
The Female equally with the Male I sing.

WALT WHITMAN

NEANDERTHALS YAWNED

Neanderthals yawned. Tutankhamen cried. Eleanor of Aquitaine belched. No doubt Murasaki Shikibu combed her hair and Askia Muhammad liked to prop up his feet. The pages of Louis XV yearned to sit down. The armies of Montezuma stubbed their toes, scratched their heads, blinked their eyes, and chewed their food; they bled when wounded and laughed when tickled. All of these people experienced their greatest pleasures and worst pains through their bodies. Throughout history how many lovers have caressed and how many war victims have writhed? Like us, they feared more than anything else the multitude of ways that the body could come to harm or even perish. Their mythologies, like ours, imagined a resurrection of the body because consciousness without it is inconceivable.

Despite the variety of cultures around the world, every human being reenacts the lives of the billions who were born and died in bodies like our own. On a metropolitan street you can see the diversity of humanity represented all around you. There are women and men and girls and boys of every size, shape, and color—different skin tones and facial contours, variations in eye shape and hair texture. Each of us has one body, and each of us comes from a culture that tells us what to do with it. In the same crowd there may be Orthodox *peyot* and skinhead pates, straightened hair of African origin and curled hair of European origin, artificial fingernails and painted toenails, tailored beards and pierced navels, false teeth and tucked necks, shaven legs and unshaven underarms, plucked eyebrows and rouged cheeks, enlarged breasts and reduced noses, calves taut in high heels and earlobes stretched by jewelry.

So much tinkering with the body exemplifies our ambivalent but creative response to it. The German film director Wim Wenders captured the joy and confusion of having a body in *Der Himmel über Berlin* (released in the United States as *Wings of Desire*), in which bodiless angels tire of their eternal voyeurism and yearn for the experience of being alive in a corporeal form. Inspired by both Rilke's poems and the director's own feelings about a divided Berlin, the script by Wenders and Peter Handke conveys a passionate longing for the sense of touch that humans take for granted. The angels hunger to grasp a pencil, caress an ear, stretch their toes, feed a cat, even to acquire blackened fingers from reading a newspaper. "Instead of forever hovering above," says one angel, "I'd like to feel there's some weight to me . . . to end my eternity and bind me to earth. At each step, each gust of wind, I'd like to be able to say, 'Now!' " After his first night of making love with a woman, he says, "I know now what no angel knows."

Daily we wallow in the luxurious physicality for which the angel yearns. The human body perceives the world through its senses, and there is no sense but touch. Through your body the world touches you. You taste chocolate and champagne when their molecules caress your tongue. You hear music when sound waves play the tympani in your ears. You smell coffee because tiny particles of it float through the air and touch the receptors in your nose. Photons enter your eyes and enable you to see the color of sunlit leaves. And there is the sense that we officially call touch, which enables you to perceive the difference against your skin of the wool of your jacket and the cotton of your shirt, or the texture of the luggage you're carrying, or sunshine on your closed eyelids.

The slow evolution that eventually created the body of *Homo sapiens*—the first animal, as far as we know, to contemplate itself— is a wonderful story. Every part of the body attests to gradual change over long periods of time. One of the Zen-like side effects of the natural sciences is the big-picture perspective of biological time. As we scurry around atwitter with the fad or crisis du jour, we forget that we are as subject to nature's laws as are the slime mold and the dodo. Even religious fundamentalists concede that, like other creatures, we select mates and reproduce our mutual charac- teristics, from the father's height or eyebrows to the mother's bone structure or skin tone, and that, in turn, our children choose mates and reproduce, further varying the pattern. What they are reluc- tant to admit is that this process has been going on for an incom- prehensibly long time. We have changed, and we are changing.

When this realization becomes a part of your everyday think- ing, you begin to see the bodies around you differently. They blur and shift before your eyes. Faces morph from one shape into an- other like those in computer-generated films. And you begin to re- alize that the human body is composed of malleable clay. Long before cosmetic surgeons approached it as a work in progress, nature was whittling and sculpting the body to adapt it for many different environments. Every inch of the human form bears the stamp of nature's restless creativity. We have prominent noses and large but- tocks. We stand upright on two legs. Our ears are situated to gather sounds and triangulate their location, and our navels mark us as placental mammals. We seem naked when contrasted with our hirsute relatives, but actually we have a great deal of hair on our bodies—much of it still clustered strategically to harbor scent.

Yet we are not content with such hard facts, no matter how im- pressive they may be. "The meaning of things," observed Antoine de Saint-Exupéry, "lies not in themselves but in our attitudes to-ward them." The human mind perceives the world in symbolic terms and never stops exercising the imagination. As a result, much of culture consists of fictions that endow natural processes with symbolic importance. There is no better example than our re-action to the talents and limitations of what we have variously called a machine for living, the temple of the soul, and our mortal coil—the human body. Every part and function of the body plays its symbolic role. In Islam the fingers of the open hand symbolize the Five Precepts. The instability of flame was represented by fire deities without feet. A haircut, a sneeze, even trimmed fingernails could mean a diminution of the vital force of life. Created in our own image, gods took their cues from the human form. Atlas car-ried the world on his shoulders. The mighty Strenua aided hu-manity with her muscular arms. Samson's hair was seen not as a protein substance produced by follicles in the scalp but as a con-duit through which God bestowed some of his own strength upon a mortal.

Like a sketch under a painting, the natural history of the body guides the composition of our faded mythologies and our flourishing preoccupations. Whatever social subsets may claim us, whatever abstract groups court our allegiance, we are also pri-mates, mammals, and vertebrates. However, before reason won its small role as explainer of phenomena, superstition assigned every part of the body its own Just-So "explanation," even such aspects as disfigurement, beauty, disease, and ugliness. The result is a sizable portion of our shared culture. With myth and art—and, more recently, with science—we have tried to answer through the body the three questions that Paul Gauguin once used as the title of a painting: *Where did we come from? What are we? Where are we going?*

THE MYSTERY OF THE VISIBLE

This book is not about my personal experiences, but one of them inspired it. Like most people with a functioning body, I take a number of activities for granted—trivial actions such as raising my head and feeling sensation in my fingertips. That is, I blithely did so until a few years ago. Then I dislocated a cervical disc and suffered ongoing neck and back pain until, after sudden paralysis in my left arm, I underwent a discectomy. My neurosurgeon told me that mine was one of the worst herniated discs of the thousand-plus on which he had operated. Pride in uniqueness didn't alleviate my pain during the two weeks I spent mostly flat on my back. To bolster his advice that I hold my head up as little as possible, the neurosurgeon provided a vivid image. He said that the human head is roughly the size and weight of a bowling ball and that the spine labors like the stem of a sunflower to carry such a burden. Not wanting the poor overworked stem to snap again, I followed the doctor's orders. I lay in bed and stared at the ceiling.

Desperate for some kind of intellectual activity, I began thinking about the human body. I found that I could rest a legal pad flat on my chest and write without having to watch the words appear, if occasionally I worked the pad to a wobbly upright position to see if my palimpsest of notes was decipherable. During much of my convalescence, I whiled away the hours scrawling free association under such headings as "Ears" and "Navel" and "Toes." Beside me on the bed the pages accumulated. Pliny's remark about King Pyrrhus's restorative toe stirred memories of Margaret Fox and the founding of American spiritualism. Houdini's cues to his assistant reminded me of Darwin's pointed ears. I could not envision Neil Armstrong's carefully arranged photo op of pressing his boot into moon soil without also seeing the row of ancient footprints that Mary Leakey unearthed at Laetoli. And every time I tried to raise my head, I remembered how much of our back pain scientists attribute to an awkward bipedalism—the once horizontal mammalian spine wrenched upward to support the aspiring head and free the greedy hands, leaving the old vertebrate nerves and their armor crowded too closely together.

Soon I realized that I had begun my next book. When I could

sit up again, I dived into research about the body. Each source led to new discoveries. In time I consulted experts. Because I am as interested in culture as in nature, I could not help noticing that many of our cherished myths about the body began with an imaginative response to its natural history. Nothing excites my imagination more than the border habitat where the two fields interbreed and form strange hybrids. For this reason *Adam's Navel* is itself something of a hybrid.

There are many ways to approach the study of the human body. Medical specialists examine the great administrative systems that govern bodily departments: skeletal, muscular, nervous, digestive, respiratory, cardiovascular, endocrine, lymphatic. Paleontologists burrow after the hard evidence of our ancestry. Sociologists, psychologists, reflexologists—every species of -ologist manages to apply a theme to the body. Athletes sculpt themselves into works of art. Books address self-image, attractiveness, sexual performance, grooming, nourishment, exercise, the sinfulness of the body, and the possibility that your soul previously inhabited a different vehicle than the one that so preoccupies you now.

Because none of these approaches covers my particular interests, I follow my own itinerary in *Adam's Navel*. I journey down the human body—male and female—from head to toe, one region at a time. I take as my model a curious style of poetry that arose in France in the mid-sixteenth century. At the instigation of an exiled poet named Clément Marot, a group of prominent writers began composing *blasons anatomiques*—poetic tributes to the individual parts of the female body. Such celebrations of body parts had their antecedents, including Petrarch's odes to the eyes of his beloved Laura in the 1300s and a salacious tribute to the breast by the later Baldassarre Olimpo da Sassoferrato. What was new was the French poets' attempt to apply this kind of admiring ode to body parts of lesser symbolic rank than the window of the soul or the nurturing bosom. Consequently the poets faced opposition. "What could be serious in the context of Laura's expressive, inspiring eyes," writes historian Nancy J. Vickers, "became absurd when applied to a random tooth or toe." Moreover, the blazons usually addressed the body part directly, a pose that sounds rather silly when speaking to the elbow. And yet, as Vickers explains, had Marot written a traditional homage to the entire female form, he would not have opened up such fertile and controversial poetic territory. Soon there were blazons in praise, *laudatio*, and counterblazons in blame, *vituperatio*, expressing the spectrum from adoration to revulsion.

In a sense *Adam's Navel* is an updated version of blazons and counterblazons, focusing largely (but not solely) on how the cultural history of the body reflects its natural history. Although I resist the kind of personification in which the *blasonneurs* indulged, I address our ambivalent regard for the vehicle—comic and tragic, divine and mundane—that carries our aspiring consciousness from cradle to grave. Our feelings about the body still range from laudatory to vituperative, as will be amply demonstrated when we zoom in for blazonlike close-ups and examine the many separate yet interdependent parts of the body. I like to think of them as the mutinous citizen describes them in *Coriolanus*:

> The kingly-crownèd head, the vigilant eye,
> The counsellor heart, the arm our soldier,
> Our steed the leg, the tongue our trumpeter. . . .

Actually I omit the counsellor heart. I keep my attention on the outward form of the body. "The great mystery of the world is not the invisible," quipped Oscar Wilde, "but the visible." Rather than exploring the hidden lungs and heart and bones, I look at the parts of the body that are visible daily to each of us—the shape of the overall face, the mouth and ears and eyes and nose, the shoulders and arms and hands, the chest and breasts, the abdomen and navel and waist, the genitals, the buttocks, the legs and feet. These areas nicely map into three distinct regions, each reflecting a different aspect of our evolutionary history: the head and face, the arms and torso, the genitals and legs. These divisions are not merely reductive intellectual constructs. They represent evidence gathered from various sources—the anatomy and physiology of contemporary humans, comparisons to our primate kin, and fossil remains of our extinct ancestors. Each of these areas of the body remembers its past differently. Each inspires different cultural responses. In this book they are discussed in three parts: "Headquarters," "The Weight of the World," and "A Leg to Stand On." These titles embody both the natural history of each particular region and a powerful metaphor inspired by it.

I chose the route downward from the head to the feet for two reasons. First, it appealed to me as narrative, a journey rather than a system. Then I remembered that each human being actually develops in the same progression. In a newborn infant, the first feature of the freshly minted body to come under the baby's control are the eye muscles. Gradually she achieves awareness of and control over the rest of her facial muscles and arrives at the popular milestone when deliberate smiling makes its appearance. Then the neck muscles come under her influence; her head no longer lolls to the side. Eventually the torso and trunk become part of the baby's sense of herself. Uncontrolled arms and finally even the legs are recruited to serve with the rest of the body, changing from clumsy appendages and chew toys into precisely manipulated hands and carefully placed feet.

Alas, many topics could not be included in this limited space. I omit, to name a few, teeth, breast augmentation, beards, weight gain and loss, the elbow and knee, and the huge topic of racism and skin color. During my journey down the body, however, certain landmarks demanded inclusion—the vigilant eye, for example. Often I realized that I was writing this book the way that I like to travel, by stopping wherever something piques my curiosity. My foray into the splendid weirdness of the human toes, or my ode to the eyebrows, I mean as examples of the black-hole density of the marvelous packed into even the humblest of topics. In both cultural and natural history, no part of the body lacks a story.

Skin Deep

> *If soul may look and body touch,*
> *Which is the more blest?*

YEATS

A single versatile membrane unites the varied regions of our bodily landscape into the appearance of what Whitman called "the form complete." The skin is considered the largest organ. No example better demonstrates how it wraps us and contains us than an ancient story about the loss of this overlooked guardian.

Greek myths tell two versions of a story about Marsyas, who either invents the double flute or picks it up after Athena invents and discards the instrument. In both accounts Marsyas creates music so mellifluous that it inspires him to challenge Apollo to a musical duel—flute versus lyre. Apollo agrees only if the winner can choose the loser's punishment. Intoxicated with the grandeur of his music, Marsyas foolishly agrees. In the first duel Marsyas and Apollo are so evenly matched in talent that the contest is declared a tie. The vain Apollo then resorts to cheating. He can play his lyre upside down, and he challenges Marsyas to match this pointless virtuosity with his flute. Marsyas cannot and loses. In one of the many insanely cruel gestures of the gods, Apollo then ties his opponent to a tree and flays him alive.

This horrific scene was disturbingly popular in Hellenistic art, and it resurfaced during the classical revival of the sixteenth and seventeenth centuries. Later Giovanni Stefano Danedi painted an almost lyrical version, and Titian wallowed in the gore, portraying Marsyas suspended by his feet as Apollo skins him. One representation that brings home the true horror is an elegant neo-classical marble by Antonio Corradini, a piece of royal garden sculpture calmly paired with the same artist's romantic *Zephyr and Flora*. Both are now in the main domed lobby of the Victoria and Albert Museum in London. The beardless, handsome, laurel-crowned Apollo casually wields a bone-handled knife, pushing and guiding it with his forefinger extended along the back of the blade. He is cutting off the skin of Marsyas's right thigh; you can see it bunching up in front of the knife.

In the first century B.C.E., Ovid vividly described the gory scene that Corradini later distilled into bloodless marble:

> . . . *and as he screamed*
> *Apollo stripped his skin; the whole of him*
> *Was one huge wound, blood streaming everywhere,*
> *Sinews laid bare, veins naked, quivering*
> *And pulsing. You could count his twitching guts,*
> *And the tissues as the light shone through his ribs.*

Why dwell upon this horror? Because it brings home to us what the skin does. Here's another example: Plastination, Gunther von Hagens's process for preserving bodies by saturating them with resins, resulted in a famous figure, a man who stands upright holding his entire skin, which has been dissected away in a single piece like the peel of an apple. He is a later incarnation of *scorticoti*, the anatomical figures prepared in the eighteenth century by peeling back the skin to reveal the subcutaneous strata. The plastinated man is as skinless as the vanquished Marsyas. Tendons, ligaments, the cabled muscles of the arms and legs; the dangling testicles and penis, the bottle-cap ring of the navel—all are revealed, pink and monstrous and natural, in as sobering an image as a *vanitas* skull grinning over a feast in a Dutch still life. Worst but most instructive of all, the figure holds aloft like a beige over-coat his ghastly shed skin. "Man is only man on the surface," wrote Paul Valéry. "Lift the skin, dissect: here the machineries begin. Then you lose yourself in an unfathomable substance, alien to everything you know and yet of the essence." To appreciate something, it helps to mentally remove it and examine the conse-quences; we need the sad tale of Marsyas and the pose of the skinned man to help us face our alien essence.

Now that we have an idea of where we would be without skin, let's look at what our wrapping does. Roughly 16 percent of our weight is skin, with the body devoting from 5 to 8 percent of its metabolism to its maintenance. It must be valuable. When we gaze upon a naked body—the face of a child, the body of a lover—it is the rolling landscape of the skin that we see. It de-fines us and protects us. Impressively versatile, it is tight across the palm, thick on the sole, flexible in the eyelid. It curves and swoops, folds and puckers. Metaphors abound: The skin is the soil atop the strata, the upholstery around the cushion, the peel on the fruit. Like all cells, skin cells are issued sealed orders instructing them to stop the growing process at a prearranged point. In doing so they delineate the borders of the self. Think of it: Without these cells we would not stop growing, would not find a shape. The skin is the line at which we become individual ani-mals moving self-contained through the world.

The skin, however, is not a space suit, transporting its own packaged atmosphere. "We are in the world," remarked Alfred North Whitehead, "and the world is in us." The world—from nu-trients to viruses—enters the body and becomes a part of it. The body takes in air and broccoli and steak and converts these for-eign substances into lungs and brain cells, somehow alchemically metamorphoses plant and animal flesh into our own flesh. Con-sider the number of worldly substances that enter and leave through just the skin, from ultraviolet light to perspiration. The skin is a boundary, not a wall.

Because of variations in the pigment melanin, the coloring of skin differs as much as do form and stature, from Inuit to Masai.

Many studies indicate that skin color varies as much *within* ethnic groups as *between* them. Hence the ability of paler or darker people, less distinctive and more average people, to hide their racial backgrounds and integrate further than social restrictions intend. Density levels of melanin are determined primarily by our genes. Exposure to the ultraviolet rays in sunlight, however, induce about the same intensity of tan in all populations, although we do not notice this as quickly on a dark-skinned person as we might on someone whose skin more readily shows the burn.

Tanning is of course a kind of burn. The brouhaha surrounding the acquisition of an approved degree of sunburn is yet another weirdly imaginative response to the way that the animal body naturally functions. Sunburn is a warning flag of danger. Although the sun is certainly hot with its own infrared radiation, it does not burn us the way that a hot coal would. Sunburn is actually a toxic reaction to the ultraviolet radiation in sunlight. Ultraviolet rays are shorter in length than visible light, making them imperceptible to us but also granting them more power. Exposure damages DNA and, during the next few hours, induces the production of enzymes and proteins that both encourage inflammatory cells and dilate blood vessels, producing the sensation of hot skin. Burned skin peels because the sun's attack on DNA kills the cells and the body sloughs them off. Repair begins immediately unless the burn is too severe, but acutely damaged DNA in cells can mutate into skin cancer. A good sunscreen performs two helpful tasks. Its inorganic molecules help reflect and scatter ultraviolet radiation, and its organic molecules absorb it.

Think of the billions of dollars that have been spent worldwide over the last half century or so to convince human beings that pale skin is a public admission of a boring and housebound life. With the postwar relaxation of strictures against how much of the body could be displayed, sun worshipping grew in popularity until late-twentieth-century revelations about skin cancer. For its effect on men, the two-piece bathing suit was named a year after U.S. atomic-bomb tests on Bikini Atoll in 1946, and suddenly beachgoing was sexier than the overdressed Victorians could have imagined in their changing coaches rolled axle-deep in the surf. A tanned body came to be considered sultry, tropical, sexy. "The marginalized tan races," wrote John Updike, "were credited with the sexiness that had been discounted by the Protestant ethic." Pale Europeans and European Americans were determined to partake of this myth and to prove that they had time and resources to loll on a beach and return with the badge of the good life. As early as 1951 Jean Cocteau was writing in his diary, "Odd that Americans, who despise 'people of color,' should be so eager to let the sun bake them as black as possible." There is now no question that exposure to sunlight encourages skin cancer, especially in these days of a thinning ozonosphere, and sunlamps and tanning beds have proven even more vicious. Nowadays, however, no matter what your climate, tan-in-a-bottle concoctions offer a safe alternative.

While we are on the topic of skin coloring, we should not forget that most of us are splatter-painted with freckles and moles. In a poem Dorothy Parker listed freckles among the four things—along with love, curiosity, and doubt—that she would have been better off without. But those concentrations of melanin provide charming spice to the face and body and even, in the case of such celebrities as Cindy Crawford, enhance signature features. Consider the term "beauty spot." It promotes a mole or some other blemish to an asset that somehow enriches the rest of the face, a sort of exception that proves the rule. It also indicates that the face was already considered attractive, because the general consensus would be that only pretty women have beauty spots, that unattractive women merely have moles. Some Renaissance ladies drew beauty spots on their unmarked faces. Proust's narrator says of his early encounters with Albertine, "In fact, when I saw her I noticed that she had a beauty spot, but my errant memory made it wander about her face, fixing it now in one place, now in another," like I-gor's roving hump in Mel Brooks's film *Young Frankenstein*. Naturally we cannot simply accept the variety of skin appearance. A plastic surgeon named Henry Junius Schireson reported that a densely freckled nurse sought plastic surgery because she looked "like a Mulatto" in dim light. So much riding on speckles of pigment.

It is easy to take for granted the skin's role in protection and to think only of the pleasure it provides. The network of nerves that invisibly bejewel our epidermis evolved in the animal body to respond to its needs in the material world, not to delight its owner. Obviously we are not the only creature to experience pleasure in the world through its effect on our bodies, but it is a safe bet that we have taken the idea and run with it to new extremes previously unexplored by our busy cousins. Good for us. Who could not be delighted with a world that offers its sensual self to our hands and feet and mouths?

Naturally any marvel as important and ubiquitous as our skin is well represented in symbolism. In Shamanism and other magical thinking, wearing the skin of an animal bestows upon the bearer the creature's unique traits and instinctive wisdom. Because snakes shed their skins as they grow, many primitive cultures incorporated real and symbolic skin into rituals accompanying initiations and other milestones in life. Any emblem of rebirth becomes a symbol of eternal renewal, and therefore skin is associated with resurrection and immortality. Some peoples have worn different-colored animal skins to represent two sides of human nature, or even the notion of manifestation versus the disembodied. These primitive associations fade slowly—if ever. In the first years of the twenty-first century, there is in the United States a renewed fashion in (usually fake) animal prints—the camouflaging pattern of jaguar spots on a designer purse, rugs adorned with the pleasing stripes of that graphic designer's invention, the zebra.

Because of the skin's ubiquity, its two square yards suffer the slings and arrows of fortune more than does any other part of the body. It survives our clumsiness, struggles with gravity, and deliberate violence. Abraded, bruised, even cut, it struggles to hold the insides in and the covering together. The very sensitivity that affords us so much pleasure results from a complex network of nerve endings and specialized cells that also provide the body's alarm system. Unquestionably the bad news in life is pain. The great sad irony, from our point of view, is that the body's repair mechanisms are not able to remedy every problem about which the alarms inform it. Whenever a predator's claw or a soldier's bullet or a blackberry's thorn pierces your disappointingly vulnerable armor, the body rushes medical personnel to your aid. Up to a point, if the damage is not too severe, these microscopic medics are able to knit back together the torn tissues. And the pain fades, only to be replaced by a new alarm going off elsewhere on the body.

At first glance the apparent defenselessness of our outer covering seems like a design flaw. Why aren't we armored to protect ourselves and reduce the thousand natural shocks that flesh is heir to down toward, say, a more manageable two hundred? Because the skin has many jobs to perform besides that of sheltering wall. Water, air, and nutrients—and poisons, viruses, and parasites—reach us through holes in the skin. The skin is riddled with doors for egress or ingress, and they vary enormously. The gaping mouth is large and demands constant attention, working overtime as fuel processor, poison tester, and microphone. The shutterlike anus is hidden away at the antipodes and serves mostly for export rather than import—barring a few intimate exceptions. Both are holes in the skin. The ear? A hole with a satellite dish built around it. Twin nostrils drink the essential air, greedily sucking in the atoms in profligate suspension around us. And of course these large visible holes in the skin are greatly outnumbered by the holes we can't see, such as pores.

Let us look at one substance that enters the body through the skin and one that leaves it via the same route. In the James Bond movie *Goldfinger,* one of the diverse tasks that the metal-hatted henchman Oddjob performs for Auric Goldfinger is the murder of a treacherous colleague by painting her entire body with gold paint. She suffocates. For decades the idea that a human being could die this way was dismissed as fantasy, although amphibians breathe through their skin. But recent studies indicate that the skin gets more of its oxygen from the air than scientists formerly thought. German dermatologist Markus Stücker and colleagues reported in 2002 that airborne oxygen penetrates almost ten times deeper than formerly estimated, ranging from a quarter to almost a half millimeter below the surface. This means that oxygen from the air rather than from the blood nourishes the entire epidermis and some of the underlying dermis, which is thickly populated with hair follicles and perspiration glands. In every age group, however, prohibiting oxygen intake by the skin had little effect on internal organs, so murder by the Goldfinger method is still unfeasible.

Down in the dermis alongside the hair roots, the sweat glands busily perform the essential function of perspiration. Some researchers speculate that the human body may have evolved its current hairless state in response to the evolution of sweat glands. Experimental pathologist Marc Lappé says in his book about skin, *The Body's Edge*, "Having a high density of sweat glands at the skin's surface provides a highly efficient mechanism for permitting the rapid cooling of the body's core through evaporative water loss—as long as this evaporation can proceed unimpeded by a hair coat." Lappé points out that tropical populations possess more densely packed sweat glands than do their temperate or wintry kin.

Even sweat, of course, has its cultural history. In general, Americans prefer to see it only on the beach and apply a great many concoctions to banish not only the scent but even the sight of it. In his early volume *Mythologies*, the first volley in his semiotics of everyday life, Roland Barthes analyzed two aspects of Joseph Mankiewicz's 1953 film of Shakespeare's *Julius Caesar*. Barthes points out that it employs two main symbolic gestures of the body, a schematized Roman hairstyle and sweaty faces on everyone except the intended victim. He even works in a feeble swipe against the English: "To sweat is to think—which evidently rests on the postulate, appropriate to a nation of businessmen, that thought is a violent, cataclysmic operation, of which sweat is only the most benign symptom."

Later Barthes writes, "In the whole film, there is but one man who does not sweat and who remains smooth-faced, unperturbed and water-tight: Caesar." Watertight: Now, there is something that we do *not* want to be. We are porous but not impermeable. Like the body of the earth itself, we are composed largely of water, and we must keep our allotment separate from the rest until we die and our spinning atoms rejoin the world's. This image brings us back to the mythological figure whose sad fate opened this glance at skin. Some stories claim that eventually the gods, always interfering much too late, transformed Marsyas into a river, and Ovid says that the tears of Marsyas's mourners created the river. Whatever its source, nowadays the river Marsyas empties into the original Meander in Phrygia. Swimmers still immerse their bodies in the water—bodies contained and defended by the skin that Marsyas lost in his battle with the gods.

PART ONE

Head-
quarters

CHAPTER 1

The Not-Quite-Naked Ape

> *Men are beastly! They are silly, vain, arrogant, and have hair all over their bodies.*
>
> COUNTESS CHARLOTTE,
> in Ingmar Bergman's *Smiles of a Summer Night*

THE HAIR OF THE PROPHET

Even God understands the importance of a spiffy hairdo. At least his vicar in Hollywood did. In his 1956 remake of his own earlier film *The Ten Commandments*, Cecil B. DeMille, while surpassing even the shameless vulgarity of *Samson and Delilah*, returned to the hairy symbolism of that film. When Charlton Heston's Moses goes up on the forbidden mountain to speak with God, he wears a short brown beard and a cloak that hides his hair. Afterward, when the prophet returns to his people laden with commandments, he has aged. The beard is longer and has two prominent gray swaths. The headgear has vanished, revealing a dramatic coiffure—long gray-white hair drawn back from his forehead and face and covering his ears. The blow-dried look is reminiscent of Michelangelo's God on the Sistine ceiling, or of the lord of the gods in Ingres's regal *Jupiter and Thetis*. Walking back down the mountainside, Moses now *looks* like a patriarch, as if he has been visiting an image consultant rather than conferring with a talkative shrub. When he returns to his tent, his faithful wife, Sephora, played by Yvonne De Carlo, expresses the sentiments of the viewer by greeting her husband with "Moses—your *hair*—"

THE MARK OF THE BEAST

With his depiction of Moses, Cecil B. DeMille demonstrated how powerfully symbolic hair can be. Our cultural response to this highly visible and easily varied part of the human body has ranged from the most primitive kind of magical thinking to forthright signifiers of caste and allegiance. The human brain is always eager to assign symbolic meaning. It perceives our animal attributes and promotes them to metaphor. Long ago it accomplished this conceptual upgrade for the lingering evidence of our mammalian ancestry.

"There are one hundred and ninety-three living species of monkeys and apes," wrote the English biologist Desmond Morris in his 1967 book *The Naked Ape*. "One hundred and ninety-two of them are covered with hair. The exception is a naked ape self-named *Homo sapiens*." In reality the species that dubbed itself wise is not quite naked. Most of us have hair on every area of our bodies except our lips, our nipples, the palms of our hands and soles of our feet, and certain parts of the external genitalia. Much of it, however, is either fine or sparse or both. Compared with our furry cousins, we certainly *look* naked.

In their microscopic analysis of our every trait, evolutionary biologists have not ignored hair. A glance at any fellow mammal raises questions. For example, nowadays elephants are almost hairless. This is not a problem for them—nor, presumably, an embarrassment—because they live in the tropics. Like human beings, they experience no discomfort running around stark naked in such an environment. Some of their ancestors, however, were famously woolly and lived as far north as the Arctic. In 1871, in *The Descent of Man*, Darwin remarked that even Indian elephants native to cool altitudes have more hair than do their lowland Indian brethren. "May we then infer," he asked, "that man became divested of hair from having aboriginally inhabited some tropic land?" As usual, Darwin immediately examined an objection to his own speculations. None of our fellow simians, most of which are fully as tropical as we, have lost their hair. Only humans emerge, as Desmond Morris would say, naked. One conjecture, mentioned in the Overture, is that we may have lost much of our hair in response to the evolution of sweat glands. Our relative hairlessness struck Darwin's colleague, Alfred Russel Wallace, as an objection to the theory of natural selection. He insisted that, even if our tropical ancestors gradually became less hairy, still they ought to have reevolved a fur coat during the hominid diaspora to less temperate climes. For this reason, he concluded, "man's naked skin could not have been produced by natural selection." Wallace was always looking for some aspect of human evolution that might testify to a little divine nudging; finally he settled upon the brain's complexity.

Darwin pointed out that in human beings the old mammalian body covering remains most dense "at the junction of all four limbs with the trunk," and, at least in the male, on the chest and face. Before our ancestors evolved bipedalism and stood erect, all of those areas would have been sheltered from the sun. But another objection immediately came to mind: "The crown of the head, however, offers a curious exception, for at all times it must have been one of the most exposed parts, yet it is thickly clothed with hair."

Not only are we not hairless as adults, we were not even naked in the womb. Roughly four months after conception, the human fetus grows a mustache. Fine hair forms on the upper lip and eyebrows. Gradually, over the next few weeks, it covers the entire body, and by the end of the fifth month the unborn fetus is completely hairy. It will remain so for many weeks. Biologists call this coat of hair the *lanugo*, from a Latin word for down or wooliness. Usually, but not always, our lanuginose phase ends before birth. During the last few weeks of pregnancy, the baby swallows the shed lanugo. The tiny hairs join mucus and bile and other substances to form the meconium, the baby's first bowel movement after birth.

Occasionally, like most other aspects of the human body, our genetic orders about hair provide the wrong instructions. This happens even with the lanugo. There is a rare genetic disorder called variously *congenital hypertrichosis lanuginosa* or *hypertrichosis universalis congenita*, which has been recorded in only about forty families, in places as far apart as Southeast Asia and Central America. Certain chromosomes malfunction, and individuals are born still covered with a longer version of the lanugo. One of the earliest recorded cases occurred in the Canary Islands in the mid-sixteenth century, when an infant named Petrus Gonsales (or Gonsalvus) was born with his entire body covered in long, soft hair. Rosamond Purcell describes the situation in her book *Special Cases: Natural Anomalies and Historical Monsters:* Gonsales "was placed with a kind of muted awe into the upper echelon of Renaissance society . . . an object of social fascination, a cultivated man who was somehow not a man." He married a pretty Dutchwoman. Unfortunately their furry children resembled the father more than the mother. Various artists portrayed the werewolfishly hirsute Gonsales and his normal-looking wife, sometimes alongside their hairy-faced daughters. In these works the family is arrayed in court finery, their human eyes peering calmly from monster-movie faces. Even a glance at these portraits suggests that they inspired Jean Marais's elaborate Beast makeup in Cocteau's 1946 film *La Belle et la Bête.* Apparently it was a variation on hypertrichosis that afflicted Julia Pastrana, a member of the so-called Root-Digger Indians of western Mexico. Anticipating the Elephant Man, she became a celebrated plaything of American and European society in the mid-1800s. In the case of Pastrana, neither serious medical commentators nor sideshow promoters could resist expressing the titillating frisson of horror they felt at the sight of a human being—especially a woman— entirely covered in hair.

Hairiness has always been a bestial symbol of nature gone wild. Ambroise Paré, a sixteenth-century physician, recorded in his book *On Monsters and Marvels* a child who was born "furry as a bear" because her mother was gazing upon a picture of a hairy man at the moment of conception. In the ancient Babylonian epic *Sha Naqba Imuru* (He Who Saw the Deep), the warrior king Gilgamesh terrorizes the people of Uruk. In time the goddess Aruru hears the citizens' prayers for rescue. From a pinch of clay she makes Enkidu, a rival whose job it will be to make Gilgamesh pick on someone his own size, thus distracting him from tormenting the ordinary mortals. Naturally the two brainless lugs become pals. The wild nature of Enkidu, who is raised by animals, is visible; he is "coated in hair like the god of the animals."

From the semiotics of rock music to the enduring popularity of the fable about Beauty and the Beast, the symbolic relationship between hair and our impure nature persists to this day. One of the items found in the luggage of Mohammed Atta, a leader of the Al Qaeda terrorists who crashed planes into the World Trade Center on 11 September 2001, was a four-page document with instructions for the last evening before the attack. Alongside an admonition to make an oath to die was a reminder to "shave excess hair from the body." Bestial impurities, we imagine, can be shed along with bestial signifiers.

Our blatant kinship with other animals makes us nervous. Declaring ourselves half angel, we fear that we are also half beast. It does not help that, like wolves and mice and camels, we are inescapably hairy. Mammals have different kinds of hair on different parts of the body, and we are no exception. Most of us were born without hair except on the scalp, eyebrows, and eyelashes. This differs from body hair. The fine downy hair that infants develop soon after birth is called *vellus*. At puberty and afterward, the body replaces vellus with the pigmented, coarser body hair called *terminal hair*. Puberty has little effect on the hair of the scalp, lashes, and brows, but those areas, as we all know, tend to show age dramatically later in life. (Because this book is a tour of the body by region rather than by theme, we will address the unique aspects of pubic hair in Chapter 11.)

The November 1941 issue of *Life* magazine applauded the striking hairstyle of a young actress named Veronica Lake. In his enthusiasm the author of the article went into microscopic detail: "Miss Lake has some 150,000 hairs on her head, each measuring about .0024 inches in cross-section." He then measured the various parts of Lake's now legendary peekaboo hairstyle and added the memorable detail that "her hair catches fire fairly often when she is smoking." The allure of Veronica Lake's hair was not in its quantity or diameter; the quoted number of hairs and their width are perfectly normal. Although figures vary slightly between brunettes, redheads, and blondes, each of us carries around an average of 5 million hairs, with roughly 100,000 to 150,000 of them located on the head. Yet hair frequently plays a role in such hyperbole. In *The Second Sex,* while denouncing the artificiality and idealization in standard notions of femininity, Simone de Beauvoir observes slyly of Rémy de Gourmont, the versatile French man of letters and Symbolist missionary, that he "wanted woman to wear her hair down, rippling free as brooks and prairie grasses; but it would be on Veronica Lake's hair-do and not on an unkempt mop really left to nature that one could caress the undulations of water and grain fields."

To understand the virtues of hair, it is useful to contrast mammals with other animals. The body heat of reptiles and amphibians, for example, is not regulated by internal thermostats. Therefore they raise or lower their own temperature by the simple expedient of seeking or fleeing warmth, sometimes merely by moving into or out of direct sunlight. When this method is not an option, they turn to dormancy. Swaddled in protective coverings, mammals are not slaves to climate and season—at least not quite so much as other animals are. (Hair is such a warm and cushiony wrap that long ago we learned to steal it from our fellow mammals.) Extremes of temperature tolerance—in other words, the range of external temperatures that the body's internal thermostat can combat—vary widely with different species of mammal. Not that these generalizations are inflexible rules; some primitive mammals also have body temperatures that fluctuate. Depending upon its environment, an echidna's body temperature can vary from 72 to 97 degrees Fahrenheit.

Not all mammals are completely covered with hair when fully grown. Adult whales have only a few bristles near the mouth or even lack hair entirely. As long as it remains dry, hair traps air, but in water it would be worse than useless. Therefore most aquatic mammals gradually evolved to their current hairless state and replaced the external wrap with a subcutaneous layer of insulating fat. But even aquatic mammals possess hair as embryos. Surprisingly, it was not hairiness that the Swedish systematist Linnaeus chose as the defining characteristic of the order to which he himself belonged; mammals are so named because of their milk-producing glands.

Like fingernails and skin itself—and, for that matter, like feathers and horn—hair consists largely of the protein keratin. This strong material is the primary ingredient in the horn of the rhinoceros—the reputed aphrodisiacal properties of which, possibly because of the presence of luteotrophic hormone, have endangered the creature. In fact, the word *keratin* comes from the Greek *keras*, "horn." It is a complex substance that resists the enzymes that usually dissolve proteins, a trait that also explains the clogged drains that afflict sinks and bathtubs: Keratin is insoluble in water. This resilience explains why exhumed human bodies often have no internal organs left, while the hair and much of the skin remain. Many people believe that hair continues to grow after death, but actually rigor mortis contracts the *erector pili*, the diminutive muscles that make hair stand on end—a holdover from ancestors in which such a sign indicated fear or anger—and the hair becomes more prominent. This reaction unites with the shrinking of decaying flesh to give the appearance of still-growing hair.

Hair grows from a mass of epidermal cells that lie, as if at the bottom of a well, in the cylindrical depression of the follicle. The follicle extends from the surface down into the dermis and sometimes even into subcutaneous layers. In the dermis, at the base of the follicle, connective tissues transmit blood to nourish the hair's growth. As the cells divide at the bottom of the follicle, they are channeled upward, acquiring pigment and becoming keratinized along the way. Downy hairs may last only a few months. Each follicle of terminal hair continues to produce for at least a couple of years and sometimes as long as six, but only the growing cells at the base remain alive. The rest of the cells are as inert as the outer edge of fingernails. When the breathless *Life* reporter stared at Veronica Lake's hair, he was looking, as we are looking every time we glance in a mirror, at a mass of long-dead protein.

For a dead substance, hair is very much alive in religion, the arts, and everyday life. It is important across a spectrum of religious symbolism. Long, unbound hair represents penitence in some Christian iconography. During the christening ceremony, Greek Orthodox priests bestow upon a baby the sign of the cross by cutting its hair in three places. Many Hindu deities wear a *sikha*, a braided topknot. Representations of Brahma always include his towering braids. Although ornately coiffed on the feminine side, the hermaphroditic Ardhanarisvara wears the piled braids of an ascetic on the masculine side. Even Ganesha, the elephant-headed minority leader of the Hindu pantheon, wears the braids atop his vast noggin.

One of the sillier records of military herdthink—although described within the fold as a noble expression of esprit de corps—is a photograph of at least a couple of dozen U.S. paratroopers in France in 1945. Looking like extras in *The Last of the Mohicans*, all wear Mohawk haircuts. When Fidel Castro was leading a guerrilla war against the Cuban dictator Fulgencio Batista in the late 1950s, his followers vowed to neither cut their hair nor shave until they had achieved their goal. By contrast, in the West African nation of Mali, the elaborate coiffures of traditional Songhai women exhibit a sense of history. They recall the heritage of the fabled Songhai Empire, which flourished in the Niger Valley of the Sudan until the Moors destroyed it in the late sixteenth century. On special occasions married Songhai women wear on their foreheads the *zoumbo,* a circular disc of hair and wool that dates back to the days of the empire. Wodaabe women wear a similar large topknot bunched over the forehead and shave their hairlines to make their faces seem longer. Muslim Fulanis denounce this ornamentation because it inhibits women from praying in the prescribed Islamic manner—by prostrating oneself and touching the forehead to the earth.

Modifications of the hairline are common. The American actress Rita Hayworth underwent electrolysis to raise her hairline. She suffered through numerous painful treatments at the insistence of her husband, who argued that her natural hairline revealed too little of her forehead and diminished the visual impact of her eyes. Photographs of Hayworth before the procedure reveal a beautiful young woman in no way disfigured by not conforming to a trendy ideal. Because men's hairlines naturally recede over the years, shaving them is a routine method of making an actor look aged. When he died in an automobile accident in 1955, James Dean looked decades older than he really was because his hairline had been shaved to age him for the final scenes of *Giant.* (Among the more gruesome celebrity mementos on record—although not quite rivaling Napoleon's penis—are locks of James Dean's bloodstained hair, cut that evening after the accident.)

Obviously human hair is more than a covering. Our busy little brains are not about to permit it to remain mere fur. Hair is symbolic of many things, and symbolic for many reasons. It is also a great source of visual, tactile, and olfactory pleasure for both its owner and observers. Besides symbolism growing out of the biology of hair, geographical differences in *Homo sapiens*— the minor variations, originally local and the result of isolated evolution—have evolved several distinct kinds of human hair: the straighter, darker hair adorning those of Asian or Polynesian descent; the many versions of blonde, brunette, and redhead among Caucasian groups; and the usually dark, more tightly curled hair resulting from African ancestry. Each culture has responded to its own hair and to those whose hair differs.

Often, too, scent is omitted when hair is discussed. Why? Is it too intimate a topic? Surely most of us have hugged a child and smelled tousled hair and been flooded with tenderness. Most people have leaned into a lover's hair and inhaled the very breath of intimacy. The American writer Sandra Cisneros nostalgically expresses how much that hair can represent and convey—how much more it is than mere strands of dead protein—in two sentences from the linked stories in her collection *The House on Mango Street*.

> But my mother's hair, my mother's hair, like little rosettes, like little candy circles all curly and pretty because she pinned it in little pincurls all day, sweet to put your nose into when she is holding you, holding you and you feel safe, is the warm smell of bread before you bake it, is the smell when she makes a little room for you on her side of the bed still warm with her skin, and you sleep near her, the rain outside falling and Papa snoring. The snoring, the rain, and Mama's hair that smells like bread.

BURN IT OR BURY IT

Hair derives its existence from within the body; it grows with an apparent life of its own; it can be separated from the body; and it is relatively difficult to destroy. Thanks to these aspects of its natural history, hair came to be considered inextricably linked with the life force itself. These attributes in turn led to a role in magic. Hair was not only important as it grew but was deemed valuable even after being cut. Discarded, it might be used by a bird to make a nest, an action that somehow acquired the reputation of resulting in headaches or even death for the hair's owner. One underlying theme of the Samson story was the widespread belief that to allow your hair to fall into an enemy's hands was to court disaster. Aleister Crowley, the influential twentieth-century proponent of witchcraft, secretly disposed of his own cut hair so that his foes couldn't use it in spells against him. This is the kind of primitive magic practiced by people who venerate the supposed relics of saints or by anyone who hoards a lock of a lover's hair.

Like other primates, we groom ourselves and each other. Inevitably the human imagination has run wild with this natural habit. One medieval belief apparently traces to the sparks of static electricity created when we comb our hair. Witches, and sometimes women in general, were thought to influence storms, to create lightning and even hail by combing their hair. Some ancient Romans, convinced that untimely haircuts inspired such atmospheric turmoil, waited until storms were actually in progress before flocking to the barber. Cutting hair at the wrong time of the moon, at a bad time of day, in March, on Sunday—all have been prohibited.

Witches could increase the power of a spell merely by shaking their hair during its incantation. Fortunately for the victim, however, the power of hair could work both ways. People who thought themselves bewitched could throw some of their own hair into a fire to afflict the witch with the pain of the flames. Another superstition claimed that the brighter the hair burned when thrown into a fire, the longer its owner would live. In India the Bhils once defused the magical powers of suspected witches by clipping a lock of their hair and burying it. The Tiwi, who live on Bathurst Island north of Australia, believe that the recently dead are lonely and desire to take their survivors to the grave with them. Knowing that they are in *pukimani* (spiritual danger), the survivors disguise themselves by painting their bodies and cutting off and burning their hair. A superstition in Germany led people to carry a bag of straight hair on the abdomen for three days to learn if they were the victims of spells. If the hair did not become tangled during this time, they considered themselves free of magical problems.

In his 1946 volume *Witchcraft and Black Magic*, Montague

Summers describes the murder of Nelson Rehmeyer, in Pennsylvania in 1929. Locally considered a practitioner of black magic, Rehmeyer was thought to have placed a hex on the Hess family. Informed that the only efficient way to battle a hex is with a countercharm, several men set out to procure the requisite lock of Rehmeyer's hair. Unfortunately he was reluctant to part with it, and in the ensuing struggle he was killed.

Other reasons for keeping hair clippings have arisen. "It is held by the lower orders," wrote an observer in Dublin in the 1860s, ". . . that human hair should never be burned, but should be buried, it being stated in explanation that at the resurrection the former owner of the hair will come to seek for it." James Frazer, the Scottish anthropologist, described in his huge compilation *The Golden Bough* the beliefs of old women in an Irish village "who, having ascertained from Scripture that the hairs of their heads were all numbered by the Almighty, expected to have to account for them at the day of judgement. In order to be able to do so they stuffed the severed hair away in the thatch of their cottages."

Historians consider a fifth-century-B.C.E. Zoroastrian liturgy entitled the *Vendidad* the source of many surviving Eastern notions about the magical power of hair. In it, Ahura Mazda, the supreme creative deity of Zoroastrianism, instructs Zarathustra in the proper disposal of bodily artifacts. The god tells him to transport both nail parings and hair clippings at least ten paces away from the faithful, twenty from fire, thirty from water, and fifty from the bundles of holy twigs called *baresma*. "Then," he adds with Monty Python precision,

> thou shalt dig a hole, ten fingers deep if the earth is hard, twelve fingers deep if it is soft; thou shalt take thy hair down there and thou shalt say aloud these fiend-smiting words: "Out of his pity Mazda made plants grow." Thereupon thou shalt draw three furrows with a knife of metal around the hole, or six, or nine, and thou shalt chant the Ahuna Vairya three times, or six, or nine.

The Ahuna Vairya was the most frequently uttered Zoroastrian prayer, roughly equivalent to the Lord's Prayer in Christianity. Fingernail clippings required separate rituals. Both hair and nails, united with that other potent bodily product, blood, were used by the ancient Egyptians in a potion that supposedly rendered a victim powerless to resist the magician's influence. And, in one poignant ritual, Egyptian widows buried a lock of their hair with their husband's body, presumably as a vow of continuing devotion in the afterlife.

SAMSON'S HAIR

I confess there are in Scripture Stories that do exceed the Fables of Poets, and to a captious Reader sound like Gargantua or Bevis. Search all the Legends of times past, and the fabulous conceits of these present, and 'twill be hard to find one that deserves to carry the Buckler unto Sampson. . . .

SIR THOMAS BROWNE

No story demonstrates the magical properties of hair better than the rough-hewn hero tale of Samson. Contrasting sharply with the rest of the solemn Book of Judges, the strongman's exploits resemble those of the Sumerian hero Gilgamesh and the Greek myths of Hercules. All three are solar figures. A requisite attribute of sun-related gods was hair radiating outward from the head like rays. To cut a lock of a man's hair was to steal from him his masculine vigor, as contained in the solar power of the hair rays. Samson's story also exhibits the petulant, revisionist air of the Rambo movies—a hero designed to rewrite history and reclaim lost dignity.

First an angel appears before Manoah's wife and declares, "Behold now, thou art barren, and bearest not: but thou shalt conceive, and bear a son." He then lists some prenatal guidelines. The future mother is to avoid unclean foods. Like mothers nowadays, she is advised to abstain from alcohol. And the angel points

out that after the son's birth, "no razor shall come on his head:
for the child shall be a Nazirite unto God from the womb." We
never learn the name of Samson's mother.

The Nazirites were a sect apart. They took their name from
the Hebrew word *nazir,* which meant "consecrated" or "dedicated," and they vowed abstinence and separation as proof of
their devotion to the service of God. Most chose the role themselves; few were saddled with it from birth by an angel. Being a
Nazirite was also usually a temporary role, lasting sometimes for
as brief a period as thirty days. Only three biblical characters are
on record as lifelong Nazirites—John the Baptist, the judge and
prophet Samuel, and Samson.

There were other restrictions besides the angel's list. For one
thing, it wasn't only the mother who was to be a teetotaler. In
the Book of Numbers, speaking through Moses, God warns those
contemplating the honor and burden of becoming a Nazirite that
they must abstain not only from wine but from all grape products. The most visible declaration of faith was the Nazirites'
hairiness, and therein lies the story of Samson. Because he is a
consecrated Nazirite, Samson misses a family milestone, the first
haircut. There are no doting parents watching when, as an adult,
he unwittingly receives his first trim.

In the King James Version of the Bible, the verse (Numbers
6:5) referring to Samson's hair is straightforward:

> All the days of the vow of his separation there shall
> no razor come upon his head: until the days be ful-
> filled, in that which he separateth himself unto the
> Lord, he shall be holy, and shall let the locks of the
> hair of his head grow.

Because Samson is a homicidal buffoon who rejects all of
the vows into which he was born, some scholars maintain that
he could not possibly have been a true Nazirite. The one vow
he keeps is allowing his hair to grow untrimmed. Apparently
Samson regards this external symbol as his one inescapable

covenant—and, because God provides him with supernatural strength no matter how ungodly his behavior, he must be correct.

Now let's leap forward a few thousand years, to 1949. After more than twenty years of directing and producing other types of films, Cecil B. DeMille returned to his favorite source for sex and violence—the Bible. He had already created such howlingly campy pseudobiblical fare as *King of Kings, The Sign of the Cross,* and the first version of *The Ten Commandments.* Then DeMille added to his résumé what he described as "the greatest love story ever told." *Samson and Delilah* starred Victor Mature as the doomed strongman and Hedy Lamarr as the Philistine temptress who betrays him. Mature was a large man, but he was not exactly Arnold Schwarzenegger. Indeed, *Samson and Delilah* has been described as one of the few movies in which the hero has larger breasts than the heroine. Not that it matters, because the Bible doesn't describe Samson's physique. His strength magically derives from his hair rather than from daily workouts. To emphasize this point, a remake might star, say, a longhaired Woody Allen.

Although the Legion of Decency denounced DeMille for "portraying Samson and Delilah as a morally corrupt couple yet billing it as a religious film," in this aspect he was remaining true to his source. The biblical pair hardly exemplify family values. In the film Samson spurns Delilah for what the rouged and lipsticked Philistine dismisses as a "milk-faced Danite lily," and she vows revenge. Later she employs all her siren's wiles to draw from Samson the secret of his extraordinary strength.

"Is it some herb you mix in your food," she purrs, "or some charmed oil you rub into your body?"

Three times she begs him; three times he lies. Finally he explains that his strength comes from God, the idol-less deity that the Philistines dismiss as an "invisible" god. Samson obliquely declares that the lion's and the stallion's manes are the symbol of their power, like the wool of the ram and the eagle's two "prime feathers."

Delilah finally gets the point and touches his hair. "*This* is the mark of your power. If it were shorn from your head—"

"I'd be as weak as any other man."

Because the scriptwriters were children of the twentieth century rather than the tenth century B.C.E., the cinematic Delilah expresses incredulity: "You believe that this great god of yours has given you your power through your *hair?*"

To find out if he is finally telling her the truth, Delilah drugs Samson's drink, and soon he is unconscious. She trims his locks with his own knife. (In the biblical version Delilah "called for a man, and she caused him to shave off the seven locks of his head.") It is the tonsorial equivalent of throwing kryptonite at Superman. When Samson awakens with a crew cut, he is unable to resist the waiting Philistines. In despair he cries out, "Throw your spears! The shield of my god is gone from me." Instead of killing him, the soldiers blind him with a red-hot sword.

Throughout the film Samson's hair has been growing ever longer. However, apparently neither womanizing nor killing multitudes with an ass's jawbone keeps him from shaving; he grows the beard that was required of him as a Nazirite only after he is blind and bound to a mill wheel. We know that Samson is blind because Mature keeps his eyes closed to reveal dark eye shadow. And, as he trudges in his sightless circle day after day, mocked by onlookers, his hair slowly lengthens. The movie finally returns to the point of the story: Samson discovers that his strength is returning with his hair. He doesn't reveal this new development until the fateful day when he is led to the pagan temple to provide sport for the Philistines. There, under their towering idol of Dagon, he wreaks his vengeance. The walls come tumbling down upon the film's numerous villains, and upon Delilah and the wayward Nazirite himself.

By the tribal logic of the Bible, with this mass homicide and his own suicide Samson is redeemed in the eyes of the Lord. In the film's coda, the still milk-faced Miriam says of Samson, "Men will tell his story for a thousand years." She underestimated. By the time Cecil B. DeMille got around to exploiting the story of Samson, it had been at least three thousand years since the first stirrings of the Hercules-like legend among the Hebrews. To this day it is the best-known fable about hair.

A METAPHYSICAL OPERATION

The story of Samson is about the sheer power of hair—its magical strength, its religious energy, as interpreted from its hardy nature. But the symbolic relevance of hair did not die with Samson in the temple of Dagon; it thrives today. Even Nazirite convictions survive. The only people who maintain a strictly traditional Nazirite lifestyle are the Falashas, the black Jews of Ethiopia, but the roots of the related Rastafarian movement also date all the way back to the time of Samson.

The *Ras* in Rastafarian is a title meaning "prince." Born in Ethiopia in 1892, Lij Tafari Makonnen became Haile Selassie upon his coronation as emperor in 1930. Only the year before, Marcus Garvey, the head of the Universal Negro Improvement Association, had predicted that soon a leader would arise in Ethiopia. With one of Africa's longest recorded histories, Ethiopia claims that its government traces all the way back to King Menelik I, who ruled in the eleventh century B.C.E. and who supposedly was the son of Sheba and Solomon. Believers consider Ras Tafari to have been a lineal descendant of this colorful pair. Such a genealogy would be a fulfillment of the Old Testament promise that, in due course, a Redeemer would arise from the House of David. Rastafarians have an ambivalent relationship with the Bible. For example, they consider the story of the Hebrews' years of slavery to be the deliberately corrupted and misrepresented history of the black races, but they find in the Nazirite tradition the divine rationale for dreadlocks.

Few hairstyles are as dramatic. Although they are no longer as politically charged as when they first appeared, the long, wild-looking locks are still regarded by many as a symbol of political unity and ethnic pride. Like Christianity, Rastafarianism grew out of the dreams of an oppressed people. It promised deliverance, in the form of a millennarian repatriation to "Zion" (Ethiopia); authorized the use of ganja (marijuana) as a sacramental tool of enlightenment; and offered the possibility of a group identity, not least with its signature hairstyle.

Dreadlocks were not native to Jamaica. The Rastas imported the style. In Kenya in the early 1950s, many local whites resisted the dissolution of the old British colonial government. The native Kikuyu responded with a guerrilla insurrection, the so-called Mau Mau uprising. Photographs of the Land and Freedom Army soldiers, showing their long, matted hair, were widely disseminated, including in the Rasta organ *African Opinion*. Jamaicans had been similarly oppressed by colonialism, from the time that Christopher Columbus "discovered" the island in 1494 until independence in 1962. Their new public identification with African origins alarmed the authorities. The Rastas adopted the dramatic African hairstyle as a badge of rebellion. In Jamaica at the time, as elsewhere, many parents of African descent admonished their children to deemphasize their ethnicity by straightening or otherwise processing their hair. Some youngsters who failed to do so were denounced by their parents as "natty head pickney." Mulatto children were considered fortunate if they were born with straight—sometimes actually called "good"—hair. Because of the uneasiness it inspired in political opponents, their hairstyle was named *dread-* (for "fear") *locks*. After the Rastas adopted dreadlocks as an official symbol of political solidarity, a new term entered the patois of the resistance—"Natty Dread."

As usual, the symbolic power of hair was recognized by both its wearers and their enemies. In Jamaica in the late 1960s, when the police arrested Rastafarians, they symbolically tamed them by snipping off their wild hair. Although hoodlums and others also adopted the style, blurring the symbol's political and religious significance, the popularity of dreadlocks continued to rise. The style became associated with the burgeoning black-consciousness movement of the 1970s. But it was widely recognized and accepted only after it was adopted by reggae musicians such as Bob Marley. By the early seventies, some believers were dubbing those who did not wear the style "baldheads." In time the term referred to any nonbeliever who rejected the teachings of Ras Tafari.

Like most religious ideas, the Rastafarian emphasis on long hair is subject to individual interpretation. In the early 1980s, David Hinds, head of Britain's pioneer reggae band Steel Pulse, wore a bowler hat onstage. When his hair began to acquire the shape of his headgear, he decided to grow it straight up. In time the trunk of hair rising up from his scalp sprouted dreadlocks around its base like roots. This tonsorial topiary attracted a lot of attention, but Hinds insisted that it was not merely a publicity gimmick. "My hair," he said, "is an expression of my devotion to the Rastafarian faith. . . . My hair is religious, cultural, not fashionable."

Soon after they were popularized by reggae musicians, dreadlocks were no longer limited to Rastas or even to blacks. Whites in the public eye began to adopt the style. In 1982 the trendy London salon Antenna announced the introduction of "bobtails, or white dreadlocks." Boy George of the band Culture Club wore a version of dreadlocks for a while. In the early 1980s, he explained to a reporter from a music magazine why he wore the discordant combination of Rastafarian hair and a Star of David. "I'm not at all religious," he said; "I think the dreadlocks look good and the star is a nice symbol. . . . Basically, I wear it to annoy people." Boy George certainly annoyed some people, more for his androgyny than for his hairstyle, but he also influenced people.

For the bandwagon followers as for the Jamaicans, there was also a practical aspect to the hairdo. As the owner of Antenna proclaimed, "Now the taboo of not combing one's hair and washing it every other day has been thrown strongly out of the window." Nowadays dreadlocks are relatively common among both women and men of African descent, and not a rare sight among white men and women in various countries far from the style's origins.

◉

The Rastafarians were not the only ones to react against the practice of torturously disguising one's natural curls. In Alex Haley's "autobiography" of him, Malcolm X told the story of a haircut that he later regarded as a milestone in his life. It took place years before he became a Muslim and acquired the name el-Hajj Malik el-Shabazz. In Boston in the early 1940s, fresh from Omaha, Nebraska, he was still known as Malcolm Little. After arranging for him to be fitted for a zoot suit, a friend gave young Malcolm his first "conk."

To avoid the cost of a barber, Malcolm gathered materials himself—eggs, potatoes, soap, a rubber hose with a sprayer, rubber gloves and apron, a jar of petroleum jelly, and a can of lye. A clerk at a drugstore knew the routine so well that he asked, "Going to lay on that first conk?" Malcolm grinned and proudly replied, "Right!" He didn't grin during the actual procedure. Mixed together in a quart jar, the ingredients formed a yellowish, jellylike "glop." Thanks to the lye, the jar felt hot to the touch. A friend of Malcolm's warned him that the compound burned the scalp and added, "But the longer you can stand it, the straighter the hair." The pain turned out to be excruciating. Malcolm emerged with his original color undiminished but its natural texture replaced by "this thick, smooth sheen of shining red hair . . . as straight as any white man's."

In this context the term *conk* is probably an altered version of *congolese*, a name for a hair-straightening compound produced from copal (resin) from trees in the Congo. In the United States during the 1920s, it became a popular method for straightening hair so that it would look less "Negro" and thus more acceptable to white society. Years later, Malcolm X said of his first conk, "This was my first really big step toward self-degradation: when I endured all of that pain, literally burning my flesh with lye, in order to cook my natural hair until it was limp, to have it look like a white man's hair." After becoming a Muslim, he rejected the style and wore his hair closely cropped. Frequently he denounced the popularity of conks and also of such humiliations as blond wigs on black women. Malcolm admired performers like Sidney Poitier and Lionel Hampton because they wore their natural hair. Many others succumbed to the fashion and wore conks—not least among them a wild and original performer named Richard Penniman, better known as Little Richard. In time, as the Afro and other hairstyles became expressions of pride during the later 1960s, the conk was considered a regressive curtsey to white society. Eventually it faded away. Many black women, however, still straighten their hair, including influential figures such as model Naomi Campbell and actress Halle Berry.

Like Samson's trim at the hands of Delilah, like the Rastafarian version of the Nazirite legacy, both Malcolm X's first conk and his later rejection of the style prove a remark by the Spanish novelist Julio Cortázar: "A haircut is a metaphysical operation."

THE ETIQUETTE OF DECIDUOUSNESS

If ever there was a topic likely to send male commentators into raptures of hyperbole, it is women's hair. Occasionally the poetic exaggeration simply does not work. Consider, for example, Longfellow's odd imagery in "The Saga of King Olaf":

> Not ten yoke of oxen
> Have the power to draw us
> Like a woman's hair.

The seventeenth-century English writer James Howell had employed the same sort of comparison—squared—in a letter written in 1621: "One hair of a woman can draw more than a hundred pair of oxen." Such analogies sound like advertisements for the tensile strength of women's hair. Alexander Pope expressed the same sentiment without the livestock: "Fair tresses man's imperial race ensnare, / And beauty draws us with a single hair." Another poetic description conjures equally awkward images, Stephen Foster's "I dream of Jeanie with the light brown hair, / Floating, like a vapor, on the soft summer air." Was it Jeanie or her hair that was floating? Poetic exaggeration aside, most of us consider hair appealing only when we see it in the mass and on the appropriate head. For some reason we generally regard a single hair as distasteful rather than beautiful.

The American ecologist Aldo Leopold once wrote that pine trees earn the reputation of being "evergreen" by exploiting the illusion that governments employ—overlapping terms of office. Evergreen plants lose their needles slowly, dropping dead ones as new growth emerges. This way they avoid the naked phase of their deciduous neighbors. The human head knows the same trick; if we did not constantly grow new hairs to replace those falling out, we would all be bald. The term *deciduous* refers to a falling off during a certain stage of growth, and on occasion it is applied to antlers or even to the wings of insects. Perhaps we ought to apply it to ourselves. Like cats and dogs and pine trees, we shed. We lose an average of fifty to one hundred hairs every day, the normal rate of deciduousness increased by such activities as combing. It is interesting to imagine what myths would have evolved about hair if we shared a trait common to many of our fellow mammals—hair that grows in synchronous cycles. We would shed a great deal of it at once in periodic molts.

Occasionally we leave our shed hairs in inconvenient places. Traditionally a suspicious hair on an unfaithful spouse's clothing can be as damning as lipstick on a collar. Criminals have been convicted on the evidence of hair found at the scene of a crime, especially with the growing accuracy of DNA testing. In the 1992 edition of her *Etiquette,* no less an authority than Emily Post advises on the proper response when encountering a stray hair. Immediately after admitting that there is no rule for choking on a bone and even advising that during the crisis one might forgo etiquette, Post provides guidelines for anyone encountering foreign objects in food. Ideally, one should remove the object and continue eating. "If," however, "it is such that it upsets your stomach (as a hair does to many people), leave the dish un-touched rather than embarrass your hostess in a private home." And she reassures us that, in a restaurant, it is appropriate to dis-creetly inform the server. From etiquette to "bad hair days," the ancient symbol of nature's power has been reduced to social trivia exemplifying our tamed new view of ourselves.

CHAPTER 2

Face-to-Face

I am the family face;
Flesh perishes, I live on,
Projecting trait and trace
Through time to times anon,
And leaping from place to place
Over oblivion.
The years-heired feature that can
In curve and voice and eye
Despise the human span
Of durance—that is I;
The eternal thing in man,
That heeds no call to die.

THOMAS HARDY, "Heredity"

THE FAMILY FACE

Our bestial hair surrounds a face that is equally animalian. Like other creatures' faces, ours is a busy area crowded with sophisticated information-gathering devices. There are twin instruments for the detection of visible electromagnetic radiation, two likewise paired for gathering sound waves, one (with two apertures) for analyzing particles floating in the air, and one for testing edibility and beginning the digestion of foodstuffs. The latter even doubles as the primary tool of communication in our species.

These high-tech instruments are not unique in themselves; they are merely our variation on the usual animal body plan that builds sensors just outside the command center. Yet our clan— *Homo sapiens* and our close kin among the higher primates—exhibits an interesting variation on the theme. As Desmond Morris observed, human beings have shed most of the hair that swaddles us in armor and metaphor. We are *almost* naked. But we are not the only species that has lost most of its facial hair. In the majority of higher primates, the hair gathers around the face much in the manner of our own—coming in varying degrees down the brow, around the ears, down alongside the cheeks, forming beards. The vervet of East Africa has proper whitish-gray sideburns on its black shrunken-head face. One of Jane Goodall's chimps at Gombe was so bewhiskered that she named him David Graybeard. Elderly orangutans tend to look like bearded sages, complete with the fat-Buddha posture. A step further down the phylogenetic scale, to such splendidly named Dr. Seuss primates as the slow loris and the golden potto and the aye-aye, brings us back among typically mammalian faces, fully as hairy as the rest of their body. Only among our nearer cousins do we see these friendly naked faces.

The adaptive virtue of a hairless face appears to be that it enables the cluster of information-gathering tools to also better serve as a broadcasting station—for broadcasting expressions, that is. This adaptation makes excellent evolutionary sense. As the animals with the most intricate (and most disturbingly human-like) social hierarchies, apes use subtleties of facial expression to convey subtleties of emotion. The higher primates are social and frequently violent omnivores. We need to be able to discern more about the intentions of our fellow tribe members than would, say, solitary herbivores. We watch each other's faces to learn which individuals to trust, which to fear, when to be alarmed, whether the other individual needs comfort or support, even to answer questions of identity—all the vast spectrum of social cues that glue together a busy primate society from moment to moment.

The comparatively hairless faces of primates account for the reasonably convincing ape masks in the original *Planet of the Apes* movies and the considerably more realistic efforts in Tim Burton's 2001 remake. Contrast the nuanced performances of the heroic chimpanzees in those films (performances at least as subtle as Charlton Heston's acting) with the two or three expressions available to the hairy-faced Chewbacca in *Star Wars*. The variety of movements visible on the naked face is demonstrated by the subtle motions studied by practitioners of one kind of Indian dance. There are supposedly thirty-six ways of moving the eyes and nine for just the eyelids, thirteen gestures involving tilting the head, and nine for the neck. There are six different ways of moving the cheeks, lower lip, and nose; with seven gestures the chin is even more versatile. There is a lot going on in the primate face, especially in *Homo sapiens'*, and we spend a lot of time looking at it. "You can never get tired of the human face," insists American film director Sam Mendes. "The best effect in any good film is the close-up."

Most of us are impressed with the differences between our faces as children and our faces as adults, but we would be shocked to observe the way that our faces changed before we were born. Because it encases the brain and is the headquarters for the major senses, the head is the first area of the embryo to differentiate. By birth it is more completely developed than the rest of the body, which is why development moves down the body from the head to the feet. From the moment that the spermatozoan and ovum join forces, the activity in the womb gains momentum. By the end of the first month after conception, although the embryo has nothing resembling a face, the front of the head has begun to move in that direction. What will in time become the nasal passages have appeared as two indentations high on the proto-blob that resembles the first rough blocking of a sculpture. Within another week or so, imaginative observers can perceive a face coalescing. Something resembling a flattened cartoon nose has migrated to the center of the area below the midline of what will be the face. Far away on each side of it, almost on the side of the head, are the undeveloped eyes.

Soon there is no question that this mass of cells is mysteriously shaping itself into a face. By eight weeks after conception, the creature looks almost like a human being. The nose is more nose-like; the eerie future eyes have converged toward it and even have eyelids; and there is not only a mouth but a mouth with distinct lower and upper lips. The ears are low on the head beside and even below the mouth, but the cheeks have appeared where they need to be. Curiously, at this stage the still top-heavy face, with its rounded forehead and skull, resembles the flat-featured stone heads of the monumental Olmec sculptures of Mesoamerica—minus the ornamental headgear. (One anthropologist suggests that these sculptures may represent fetal *Homo sapiens*.) By five months after conception, there is unquestionably a baby human being in the works. The mouth is smallish, the eyebrows barely developed, the ears still tiny, but the face has achieved essentially the composition and proportions it will have at birth. Evolving through a whole spectrum of artistic parallels, it now looks like Charles Schulz's round-headed characters in the early days of *Peanuts*.

When the baby emerges, the head, besides being misshapen by its traumatic squeeze through the narrow birth canal, is entirely out of proportion to the rest of the body. The eyes, already two-thirds of the size they will attain, seem huge. Prior to puberty, the most dramatic change between newborn and adult human beings is the rest of the face's growing into proportion with the prematurely large eyes. For this reason a large head and oversize eyes are almost the only feature required to make a doll look infantile, an illusion that triggers a primitive animal nurturing response in most human beings. In a 1970s essay, "A Biological Homage to Mickey Mouse," Stephen Jay Gould traced how the skinny rogue of the early Walt Disney cartoons evolved into the fat and happy shill that struts around Disneyland. With graphs and diagrams, Gould demonstrated that Mickey had slowly grown a rounder head, larger in proportion to the rest of his body, and much larger eyes—along with sleeves and gloves and baggy pants that fill out the rest of his form to a more juvenile shape. Mickey ceased to be the laughing rascal and mellowed into the squeaky uncle, foil for and victim of the other characters' misbehavior. Children and adults began to profess their love for him. He was officially cute. Gould was not wasting time in a purely academic exercise. The proportions of the infant face relate to our notions of cuteness, our fondness for equally big-eyed and big-headed puppies and kittens and chimpanzees, and even our sense of appropriately "childlike" proportions. Sociologists have found again and again that children who are born with more adult-looking features are likelier to be victims of child abuse than are their cuter fellows.

Thanks to nature's slow tinkering, its tireless puttering about in the workshop, the photographs on your mantel look the way they do. Long before Thomas Hardy wrote the poem at the head of this chapter, Shakespeare in his sonnets employed the idea of the face as a family heirloom. "Look in thy glass, and tell the face thou viewest," he begins the third sonnet, "Now is the time that face should form another. . . ." He goes on to ask who could be so absorbed in self-love as "to stop posterity?" And so his final point: "But if thou live, remember'd not to be, / Die single, and thine image dies with thee." It is a powerful argument. What parent hasn't proudly smiled when told, "She has your eyes"? Surely all parents look into the faces of their children and think at times, *How can it be that we have created you?*

The family face is a perfect example of and symbol for evolution. Combining aspects of both parents, each face performs its own unpredictable variation on the family constants and subtly enhances or undermines what had been the recognized facial legacy. Nor does this resemblance go only as far back as the black-and-white photos of your great-grandparents. The face we love to look at also remembers our heritage as primates, mammals, and vertebrates. Our genetic heritage in our family as in our species acts, in Proust's phrase, as "the invisible sculptor whose chisel repeats its work upon successive generations." The face serves as identifier and broadcast medium for that intersection of the past and the future, of biology and culture, of body and mind, that we blithely call the individual human being.

THE FACE ON MARS

We are symbols, and inhabit symbols.

RALPH WALDO EMERSON

We are surrounded by the naked simian faces of our fellow human beings, but most of us also notice facelike *patterns*—in clouds, rugs, weathered tree trunks, curled electrical cords. Few of us worry about the world's distracting tendency to peer back at us, because the illusion seems both benign and widespread. The French microscopist Louis Joblot described in 1718 observing a facelike amoeba that resembled a "perfect Mask" with "a singular coiffure," although he admitted that it also had a tail and six legs. Jeffrey Hood, an American painter, once wrote about a motel "where the circuitous patterns on the wood panelling created dozens of facial images in what was supposed to be the faux knotholes of the equally faux wood. I remember another abode where the patterned linoleum just in front of the toilet gave the distinct image of a face when viewed from the seated position."

In July 1976, NASA's *Viking 1* spacecraft went into orbit around Mars and began broadcasting the first close-up photographs of the red planet's surface. Most revealed what astronomers had predicted—an arid landscape reminiscent of the American Southwest. Some of the vistas of scattered rocks and crumbled mesa lack only sagebrush and pickup trucks. But one NASA photo, designated 035A72, presented an image that human brains were eager to misinterpret. It shows a mesa in the Cydonia region, sidelit by low-angled sunlight. The lighting creates shadows in the appropriate places to be interpreted as an eye, the side of a nose, and a mouth; the other side of the "face" is entirely in shadow. Surrounding the bodiless head is a hairlike shape, or possibly some kind of headgear; in fact, this mile-long rock formation resembles a depressed Prince Valiant glaring up at the sky. And nearby is a large impact crater, as if the gods threw a pie at the face and barely missed.

The face was discovered by a member of the imaging team that was examining photomosaics for potential landing sites for *Viking 2*. He showed the facelike image to a colleague, who later casually remarked to the press, "Isn't it peculiar what tricks of lighting and shadow can do?" Most people found the photo amusing but not particularly newsworthy. It was NASA itself that first distributed copies with the light-hearted caption "Face on Mars?" Just as it took years for the government's own account of the nonevent at Roswell to mutate into a supposed cover-up and capture the fevered imaginations of the right people, so did the Martian face only gradually seep into what passes for thinking among the UFO crowd. Not until the *Weekly World News* featured the Face on Mars in a 1984 story did the public at large seem to notice.

Since then, however, the face has inspired a barroom brawl of theories, many of which have the hardy survival virtue of requiring no evidence for their sustenance. Most claim that the face was deliberately constructed by intelligent beings. If "obviously" this is an artificial monument, who went to such trouble? Was it created by a Martian civilization that had flourished in the past—or that even now might lurk in underground cities? Remember: These would have to be Martians who looked remarkably like their Earthly neighbors. Or was it built by creatures from another solar system, who used Mars as a handy billboard on which to sculpt a humanlike countenance? Unfortunately, proponents of this theory never explain *why* spacefaring creatures would choose to taunt galactic rubes with their own physiognomy. Some argue that the landmark's supposedly eerie similarity to a human face indicates that it was created by our own species. Does this mean that our descendants are traveling back here from the future and cryptically playing in the interplanetary sand instead of leaving helpful advice about ozone depletion and species loss?

In 1998 the Mars Global Surveyor sent back to Earth another image of the face, Photo 22003, that showed the scene in ten times greater resolution than the first picture. It didn't happen to be sidelit in low light and therefore bore no resemblance whatsoever to a face. Will everyone who is surprised by this development please stand up? Since 1998 photos with even finer resolution have appeared, proving definitively that the mound is not an artificial construction and that it resembles a face only from one angle and in a certain slant of light. Yet the stories are still out there. One fringe stargazer, Richard C. Hoagland, is still promoting the idea of the face and the civilization it represents, and still selling his goofy book *The Monuments of Mars: A City on the Edge of Forever.*

As the Martian story indicates, we require very little encouragement to imagine that we see a face. The melancholy Prince Valiant is not even the only astronomical face in the cosmos. In his thoughtful and amusing book *Captured by Aliens: The Search for Life and Truth in a Very Large Universe,* Joel Achenbach says that there is a volcano on Mars that looks like Ted Kennedy. And, as you can see in many photos, the floor of Galle Crater on Mars features a lopsided, rather evil-looking smiley face.

It seems fitting that the most tiresome symbol of the last few decades would show up in space, as if some terminally happy maniac had leased the largest possible sign on which to promote inanity. Psychologically akin to the nonexistent Face on Mars, the smiley partakes of our same innate urge to see a face. It consists entirely of two dots for eyes and a curving line for a mouth. Pushed to the very limit of simplicity—in other words, to the brain's limit of recognition factors—the smiley face even beats, for sheer graphic shorthand, Milton Glaser's famous I♥NY, which depends upon a consensus about both symbol and letters.

The yellow smiley face is no longer ubiquitous, but there are plenty of other simplified faces around. One such group has even earned itself the vile portmanteau label *emoticons*. Practically everybody has received e-mail messages signed with a certain symbol—the familiar, even kitschy :) The preceding sentence had to forgo terminal punctuation because, with a period after the parenthesis, the extra dot below the upcurved lips ruined the illusion. We are so disposed to see faces everywhere that we can create the illusion of one with flyspeck-size punctuation. It doesn't even matter that the tiny pseudoface is *sideways*. We still get it; our eyes perceive the slightest hint of a facelike composition. We can even type a semicolon instead of a colon and immediately see a *winking* eye. It's ridiculous.

Why do we sometimes see faces that aren't really there? Because evolution has established just how important it is for us to perceive the faces—and expressions thereon—of our fellow human beings. Our brains perform this task so quickly all day long that we don't even notice. Imagine the questions it asks itself: *Acquaintance or stranger? Friend or enemy? Angry or happy?* We must know the answers to these questions, and we must know them as quickly as possible. So why do we sometimes see faces before there are actually faces? Because our hair-trigger identification system sometimes misfires in its eagerness to perform its job. Unwilling to wait for unambiguous data, because not recognizing a real face could have worse consequences than mistakenly imagining one, the brain exclaims, *Hey, there's a face!* and a moment later admits sheepishly, *Oops, my mistake.* This better-safe-than-sorry mechanism shows up in many of our responses to all sorts of stimuli. Predators tend to shoot first and ask questions later, and creatures that analyze all the data before reacting soon find themselves extinct. In fact, one duty of reason is to serve an inhibitory function, riding guard on such instinctive urges, making sure that we don't behave in paleolithic ways unnecessary and even counterproductive in the culture we have since evolved.

The eye and brain never stop seeking patterns in the world around them. What are constellations but chance configurations of stars that, from our particular location in the universe, happen to resemble the cruciform pattern of a flying swan or the sinuous twists of a scorpion? Such resemblance is slight at best, but enough for our pattern-hungry brains to seize upon and invent stories about. Although the brain does like to process information in this kind of symbolic shorthand, our tendency to see faces is a special circumstance. "Newborns are strongly attracted to faces, whether real or schematic," writes neurobiologist Lise Eliot. "Faces satisfy many requirements of their limited acuity; the bright eyes and dark mouth provide a lot of contrast, while the hairline provides a distinctive external frame to stimulate the baby's peripheral vision." This innate orientation is particularly impressive considering the poor vision of newborns. As proven again and again in tests, within an hour after birth, infants respond more—by moving their eyes and even their heads—to a drawing of a face than to a drawing in which the features may be present but are not arranged into a face shape. Neurologists now understand that the brain is hardwired to recognize human facial patterns, although of course it requires experience to apply the subtleties of this inborn predisposition to the particular faces in an individual's life. Many experiments confirm that we recognize human faces more easily than we do chimp faces, and presumably chimps have the same inborn bias. We also recognize individuals within our own ethnic group more easily than those from outside it. A newborn may recognize her mother or father within a day or two, but it will be a couple of months before she can discriminate among the many other faces around her. Only at this point, as the cerebral cortex matures, can infants fully perceive stationary faces rather than the slowly moving ones that they are born preferring. As long ago as 1972, scientists found the area in the cerebral cortex of monkeys that responds to faces. It produces an electrophysiological response that can be monitored, like so many other brain functions. Cells in the monkey's cortex react to a spectrum of faces and facelike shapes, while ignoring stimuli that don't fit the pattern. As always, it is depressing to think about the horrors that laboratory animals underwent to offer up these intimate details, even as it is exciting to learn more about our own factory-issue hardware.

The clearly delineated areas of the brain devoted to facial recognition explain why otherwise functional adults sometimes lose their ability to recognize even their own children. To describe this sad dysfunction, somebody united a Greek word for face, *prosopon*, and a term for lack of knowledge, *agnosia* (which Thomas Huxley also used as a root when he coined the term *agnostic*) into *prosopagnosia*. In his book *The Face*, Daniel McNeill tells the story of a man who was talking with his physical therapist when he experienced a painless stroke. Neither he nor the therapist realized this until the patient exclaimed, "But miss, what is happening to me? I can't recognize you anymore." Prosopagnosiacs have been known to lose the ability to identify their spouses or their children. Some can't recognize expression or even gender.

Although our cousins are programmed at birth to perceive faces in much the way that we are—just as most other animals come into the world ready to perceive their own kind—there is at least one way in which human beings respond uniquely to faces. We represent them in artworks. For three years beginning in 1912, until her work was halted by the First World War, the Russian scientist Nadjeta Kohts watched a juvenile chimpanzee named Yoni create pictures. At first she held Yoni on her lap and encouraged him to doodle on paper with a pencil; later she added color. In experiments with Yoni and other chimps, Kohts concluded that young chimpanzees have a rudimentary sense of form and shape that resembles that of human children. And, as primatologist Frans de Waal has pointed out, "chimpanzees are physically stronger and have better motor control than young children. Their paintings immediately strike us as forceful statements, whereas a young child's art tends to look tentative and hesitant." By the late 1950s, the chimp Congo was creating beautiful paintings that even earned an exhibition. Later work with capuchin monkeys indicated that they have less artistic bent.

Neither apes nor monkeys cross the border to representational art. However, because children begin to create representational art at an early stage, a universal early image is that of a face. Indeed, Desmond Morris points out that in many cases both human and chimpanzee infants proceed with such drawings in a similar progression. They daub paint on a page at random and then begin to concentrate their daubs in one area. Like humans, many chimps progress to some kind of circular or ovoid closed form and may even concentrate their daubs inside the circle. Soon, however, the young of the two species part company. Human infants invariably progress to simple representations of faces inside the circles. Chimpanzees, on the other hand, even if they have a basic appreciation for color and composition, never make the conceptual leap to representational art. They remain abstract expressionists. Meanwhile, the human children begin a lifetime of drawing and doodling and seeking a shape for which the very brain itself seems always hungry—the face of another, a gaze to mirror our own. "I live in the facial expression of the other," wrote Maurice Merleau-Ponty in *The Primacy of Perception*, "as I feel him living in mine. . . ." We yearn for another human being to meet our gaze and nod and perhaps even smile, a wordless gesture that declares, *Yes. I see you. We are fellow human beings. We are together.*

CHAPTER 3

The Vigilant Eye

The fact is, looking ourselves in the eyes is no easy matter.

ITALO CALVINO

FLESH BECOMES A MIRROR

Our busy naked faces are indeed equipped with ancient technology for receiving and broadcasting information, but usually our gaze is drawn first to one particular area—the eyes. In crowds we may peek at strangers' hands or clothes or hair, but we avoid meeting their eyes because the gaze is so powerful. The eyes rule the face. We refer to these organs all day long without noticing.

"Do we see eye to eye on this?"

"Don't roll your eyes at me."

"Keep an eye on her."

"I saw you eyeing him."

These are figures of speech but also reasonably accurate expressions of the act of using our eyes. The brain's dependence upon these small organs inspires more intricate symbolic notions, as it promotes the act of seeing to a metaphorical level. In Islam the spiritual core of the Absolute Intellect is called the eye of the heart. The ancient Sumerians surrounded altars with gypsum figurines whose huge staring eyes represented the awe of worshippers. St. Matthew called the eye "the light of the body" and advised Christians to pluck out their right eye if it offended. The architecture of sacred gathering places includes the concept of a spiritually aspiring "eye" at the top of a dome. For that crotchety pagan Henry Thoreau, Walden Pond was the eye of God. His neighbor, Ralph Waldo Emerson, took ocular symbolism in another direction. In his essay "Nature," Emerson wrote, "Standing on the bare ground . . . all mean egotism vanishes. I become a transparent eyeball; I am nothing; I see all; the currents of the Universal Being circulate through me; I am part and parcel of God."

Even silly eye myths reflect the organ's importance. In Greek mythology the three Old Women who have never been young, the Graeae, are rich in vowels but poor in teeth and eyes. They share one of each between them. As if appointed by a government bureaucracy, the Graeae wind up with the job for which they are least qualified—security guard. They help protect their sisters, the Gorgons. When Perseus comes gunning for Medusa, the Graeae take turns keeping watch, sharing their one eye. In an easy setup worthy of a James Bond movie, Perseus simply steals the one good eye and sneaks past the now blind, one-toothed guards. In another myth starring the same characters, Perseus adds insult to injury by stealing the eye *and* the tooth.

Eyes are ubiquitous in the arts. In Charlotte Perkins Gilman's feminist classic *The Yellow Wallpaper*, the narrator's first hint that there are figures in the wallpaper is the vague sense that the pattern of stains and rips is beginning to resemble two bulbous eyes staring at her. Cartoonists employ all kinds of symbolic shorthand when portraying eyes. In Max Beerbohm's caricatures, politicians almost always have closed eyes, and apparently it was *New Yorker* pioneer Peter Arno who invented the habit of conveying drunkenness by substituting Xs for the eyes. Cinema has made use of the eye in countless visual metaphors. Think of the way that Trevor Howard removes a mote from the eye of Celia Johnson in *Brief Encounter*, helping her begin to see her life in a new way. In *Psycho*, Alfred Hitchcock slowly fades from the round pupil and iris of Janet Leigh's wide-open dying eye to her blood running down the circular shower drain.

It has been a long, long time since the human eye was merely an organ of sight. Among the visible body parts, its only rival as a figure of speech—which can be an informal measurement of its importance in our thinking—is the hand. The brain does not permit anything to remain purely itself; it places everything in a symbolic position in the allegorical melodrama that is the mental life of *Homo sapiens*. No doubt other animals have misconceptions based upon their specialized perception of the world, but if they carry with them a labeled Looking-Glass map of reality that rivals ours, they politely keep it to themselves. We do not. The ancient Greeks, for example, considered the eye a partner in vision, casting a pale ray of its own, like a bat's navigational sonar or like a questioner demanding a reply. Now we know that the eye waits and processes the electromagnetic signals that come to it. In the case of vision, the subjective brain uses the objective eye as its *camera lucida*.

"I am a camera with its shutter open, quite passive, recording, not thinking," says the narrator of Christopher Isherwood's novel *Goodbye to Berlin*. Isherwood is talking about observation as the raw material of art, not about the eye itself. Yet the eye's role is the only truly objective one in this process, because the interpretation of visual information—even the decision of where to aim our gaze—is made in the brain. And yet the process begins with something tangible entering our eyes. Because they admit only light, we think of the eyes as windows rather than doors, but the visual part of the electromagnetic spectrum is physical. Photons reach the end of their eight-and-a-half-minute flight from the sun (at least for the tiny percentage of the sun's extravagance that this planet catches) and enter our eyes. Light is the ancient fire of the cosmos, what the English physicist Frank Close calls "the radiant glory of the electromagnetic force," literally entering our eyes and igniting tiny electrochemical fires in the brain, beginning the process of creating pictures of the world around us.

Nothing more powerfully reminds us of the interconnected-

ness of the senses, the hierarchy in the brain that obeys one
voice instead of another, than the dominance of vision. The eye
is king. We must demote it to emphasize another sense. "Stephen
closed his eyes," writes James Joyce early in *Ulysses*, "to hear his
boots crush crackling wrack and shells." Descartes summed up
the standard opinion of vision's powerful lobby in the brain:
"The whole conduct of life depends on our senses, among which
vision is the noblest and most universal." Before him was
Leonardo: "The eye, which is called the window of the soul, is
the principal means by which the central sense can most completely and abundantly appreciate the infinite works of nature."

Just as he tinkered with everything else, Marcel Proust modifies this idea of the eye as the window of the soul: "the eyes,
those features in which the flesh becomes a mirror and gives us
the illusion of enabling us, more than through the other parts of
the body, to approach the soul." What we forget is that this notion is indeed an illusion. As a result of our dependance upon the
visual, we famously fall for surfaces. Although he declares flatly
that "Man cannot naturally Percieve but through his natural or
bodily organs," William Blake also worries that we believe a lie
when we see with, not thro', the eye. Sometimes we save ourselves from deception by realizing in time what Tallulah
Bankhead once said about a Maeterlinck play: "There is less in
this than meets the eye."

You cannot glimpse the other person's soul, but when you
look into the eyes of another, you can genuinely see yourself.
The word *pupil* comes from the Latin *pupilla*, meaning "little
doll"—which is why the same word also describes a student—and
refers to the little figure of yourself you see reflected in the eye of
another. "My face in thine eye," says John Donne to his beloved,
"thine in mine appears. . . ." In Roald Dahl's children's novel
Danny the Champion of the World, the eponymous narrator describes another character: "The eyes were large and ox-like, and
they were so near to me that I could see my own face reflected
upside down in the center of each." Dahl's great virtue as a writer
was the freedom and audacity of his imagination, but sometimes

he failed to anchor it with observation—unless Danny is seeing all the way through the cornea of the other person's eye to his retina. The rays of light passing through the lens form an upside-down image on the retina at the back of the eye, but the reflection in the pupil is not vertically inverted. Until Leonardo discovered otherwise, by dissecting cadavers, Renaissance authorities insisted that vision occurs within the lens itself. Leonardo also perceived that the image upon the retina would be upside down and speculated on how this inversion might be corrected in our perception.

Sappho tells one of her lovers, "I ask you to reveal the naked beauty of your eyes." Even though they do not always reflect the soul, for those of us untrained in politics or other professional lying, the eyes reveal so much that at times we hide them. "Turn away thine eyes from me," sighs a hot-blooded lover in the Song of Solomon, "for they have overcome me." Different occupations protect the delicate face with safety goggles and fencing guards and diving masks, but these items disturb viewers only when they hide the expression in the eyes. Sunglasses were not invented to disguise the eyes—for centuries Eskimos have crafted glare goggles out of bone or wood with narrow view slits—but quickly they earned the nickname "cheaters." Sometimes police officers and gang members enhance their carefully projected image of invincibility with dark or mirror-lens eyewear that prevents their opponents from glimpsing even a flicker of emotion. You can see the strangeness of rows of hidden eyes in old photographs of the audience at 3-D movies in the 1950s or in photographs of the witnesses patiently—and with such heartbreaking innocence—awaiting an atomic bomb test in the 1940s. When there is no possible danger, the absence of expression becomes merely intriguing. The glitterati understand the allure of hidden eyes. Sunglasses are de rigueur for Hollywood shindigs, Cannes pool wading, and your general paparazzi dodging. They were a trademark of General Douglas MacArthur and remain so for celebrities such as actor Jack Nicholson or fashion guru Karl Lagerfeld.

A truly fierce expression requires the cooperation of the eyes.

Staring and glaring send powerful messages and even lead us to say, "If looks could kill . . ." Over the centuries countless people have thought that looks *can* kill. Around the world there are many variations on the idea of the evil eye. In Chapter 1, we encountered Zoroastrian rules about the disposal of hair; the Avesta, a similarly number-obsessed Iranian text, attributes 99,999 ailments to the eye of the evil god Ahriman. Coleridge's sailors cursed the Ancient Mariner with their eyes. In Greek mythology the Gorgon Medusa turned to stone all who gazed into her eyes. Perseus had to watch her reflection in his shield in order to approach close enough to behead her. (We will return to Medusa in, of all places, the chapter on the genitals.) In his *Historia Naturalis*, from the first century c.e., Pliny the Elder tells the story, which he attributes to the Greek grammarian Isigonus, of people "among the Triballi and the Illyrians, who cast spells by a glance and kill those they stare at for a longer time, especially if it is with a look of anger." Pliny mentions that adults were more susceptible to the evil eye, but apparently it did not occur to him that this might be because younger people had not had time to absorb as many years of superstition. Incidentally, not everyone agrees with Pliny on the matter of youthful resistance to a malevolent gaze. Romanian peasants used to hide mirrors from their toddlers so that the children would not inadvertently give *themselves* the evil eye.

Because we simultaneously focus two eyes on an object, employing a sophisticated method of determining distance and perceiving depth, we call our vision *binocular*. This trait is unique to vertebrates, and almost every vertebrate that can see at all employs it. With two images entering the brain through a pair of eyes, there occurs something called retinal disparity—the slightly differing views of the object forming on the wall of the retina at the back of each eye. The closer the object, the greater the disparity between the two images, just as objects side by side (on the same plane) in a photograph may be in focus while those in the foreground or background may not. Like a stereopticon overlapping a pair of photographs, the brain unites the two views into a single image with apparent depth.

Binocular depth perception is a great idea, but in case you decide to make an animal, it is not the only issue to consider when placing a pair of eyes. There is also the question of panoramic field: how far a creature can see around itself without moving its head. You can measure the range of your own eyes' panoramic field with the following demonstration of peripheral vision. Stare ahead at a fixed point; hold your arms straight out at your sides; now move your arms backward until you can no longer see them. Creatures with eyes in the front have a greater overlapping binocular field, but those with eyes on the sides can see more of the area around them. Because they pursue other creatures, predators are more likely to have eyes focusing toward the front, just as their prey tend to have eyes cautiously viewing a larger area around them. Owls and other raptors face forward, whereas mockingbirds and starlings see a wider range; cats face forward, while rabbits can see almost all the way around. Like other omnivorous primates, we have eyes firmly in the front.

Biologists theorize that contemporary primates' last common ancestor must have lived in trees because primates' paws resemble those of other tree-dwelling creatures. However, the attempt to expand the arboreal theory of evolution to other features of the primate body have not always worked as well. Many scientists have now abandoned the notion that the primate style of eyes evolved as an adaptive response to the need to leap about in trees. The English anthropologist Matt Cartmill points out that "although bringing the eyes together at the front of the head does increase the arc of stereoscopic vision," the resulting binocular overlap reduces "the distance over which stereoscopic depth perception can operate." Nor, he insists, do other arboreal mammals tend to have larger brains or decreased olfactory acuity, those other primate characteristics often tagged as the result of aboreality. Such changes in our views about seeing reflect the ongoing revision that keeps science lively.

No matter how the demands of past environments may have shaped our eyes and their location on our heads, the act of seeing is quite a talent, and we are rightly impressed with it. In what is called the "argument from design," many people—including but not limited to biblical literalists—contend that the complexity of nature is the best evidence for the existence of a creative deity. Perhaps the most famous expression of this idea is William Paley's argument about the eye in his elegant and readable 1802 book *Natural Theology, or Evidences of the Existence and Attributes of the Deity Collected from the Appearances of Nature*. Paley begins the book with his now famous watchmaker analogy. He explains the differences between natural and manufactured objects, arguing that if he found a watch on a heath, he would assume from its complexity that it had been created by an intelligent designer. He then compares the eye to a telescope and says flatly that "there is precisely the same proof that the eye was made for vision, as there is that the telescope was made for assisting it." He expands upon this idea at great length.

In his influential 1986 book *The Blind Watchmaker*, English biologist Richard Dawkins refuted both Paley's intention and his examples. "The analogy between telescope and eyes, between watch and living organism, is false," insists Dawkins. "Natural selection, the blind, unconscious, automatic process which Darwin discovered, and which we now know is the explanation for the existence and apparently purposeful form of all life, has no purpose in mind. It has no mind and no mind's eye. It does not plan for the future." Dawkins then documents the ways in which the human eye, for all its intricacies, could have evolved through a succession of subtle changes. He goes into even greater detail in a later book, *Climbing Mount Improbable*. In the vast expanse of geological time, countless generations are available for such research and development.

Dawkins examines some of the holes in this argument of creationists who seem convinced that God needs their help in proving his existence. One of their standard "proofs" of how the eye defies the notion of evolution is the claim that every single aspect of it as it exists now must work perfectly for it to function at all. Cornea, iris, retina, lens, aqueous humor—they work together. How, demand the promoters of design, could a "random" nature superintend the simultaneous interactions required for such harmony? In reply Dawkins points out that the eye does not have to work perfectly to be useful. His simplest example: Without their eyeglasses, most people with impaired vision can still see to avoid obstacles and perform most of the requirements of life. Limited visual acuity is simply not the same as blindness. Even a rudimentary response to a shadow blocking the sun is useful to an animal trying to avoid predators. Every incremental advance in such light-responsive technology is an improvement in the creature's defense system and in its increasingly sophisticated response to the world around it. As science writer Daniel McNeill remarks, beginning with the rudimentary photoreceptors that guide protozoa, in animal eyes "the evolutionary stages are out in plain view." Distinctly different but quite complex organs of vision have evolved numerous times independently among arthropods, invertebrates, and vertebrates.

Certainly there are other views besides those of Paley and Dawkins. Evolution by natural selection, which is as demonstrable and documented as gravity, does not preclude or disprove the existence of a guiding divinity, any more than the eye's complexity somehow validates ancient Hebrew fables. Science helps us figure out how the world works; it does not tell us if anyone set it in motion or adjusts the cogs daily or listens to our prayers. But for many people, in its focused attention to the hard realities around us, science is as respectful and admiring of Creation as is painting or religion or poetry. What science has learned about the eye may not confirm the existence of God. Yet it certainly testifies to the glory of the universe, as we run around peering eagerly about like every other creature—except that, so far as we know, we are the only animal dazzled by the splendor.

SUPERCILIOUS

Charles Dickens excelled at outrageous gestures that distill character into imagery, and with a single trait he enhanced the slimy unreality of Uriah Heep: The 'umble rogue has no eyebrows. A lack of hair above the eyes is unusual but not dangerous. We could survive perfectly well without our eyes' furry awnings, or for that matter without every other bit of hair on the body, but we would miss some essential tasks that they perform. What you would notice first would be the absence of a usually overlooked assignment of the brows—keeping dust and perspiration out of the eyes. At one time scientists agreed with Cicero that this was the eyebrows' primary function, but studies of expression have since replaced the awning theory.

The eyebrows' most important job seems to be that of playing supporting actor to the eyes' starring role in expressing emotion. Studies of facial expression reveal that the eyebrow is crucial in quickly projecting mood or performing other sorts of communication. Most prominent when arced in surprise or thoughtfully drawn together, the brows also support such varied actions as shrugging or expressing admiration. Cartoonists have long known and exploited the way that eyebrows add an extra outline to the eyes, attracting more attention to the orbs below. In older animated cartoons, characters' eyebrows jump above their faces to express surprise or dismay, a graphic shorthand as quickly understood as eyes jutting out of their sockets or steam shooting from ears. Discussing the IgNobel Prizes awarded each year to both honor and lampoon outrageous scientific research (such as the navel-lint survey described in Chapter 10), the 2002 organizer Mark Abrahams explained, "These are all research projects that raise eyebrows. Some raise your eyebrows so much you can damage your face."

Scientists call this distinctive action of the eyebrow—on humans or other primates—the *brow flash*. The expression, or aspect of an expression, shows up in many situations, but it is especially noticeable during the act of greeting a newcomer. Darwin included observations of baboon and other primate eyebrow gestures in his seminal 1871 volume *The Expression of the Emotions in Man and Animals*. A loving and indulgent father, he found nothing contradictory in spending his days making notes on the growth and development of his various children and then contrasting their actions with similar behavior among infant primates at the zoo. His afternoons there paid off. More systematic field studies of primates did not begin until the second half of the twentieth century, but ever since then scientists have been recording this kind of behavior. Robert M. Sapolsky, a primatologist and neurobiologist who studied baboons in the wild for years, described a randy young male baboon passing two females on a path and performing a brow flash, "the male baboon equivalent of tipping his hat."

As caricatures and party masks indicate, the iconic face of Groucho Marx owes as much to his larger-than-life eyebrows as to his greasepaint mustache. He could imply more naughty behavior with an eyebrow wiggle than many actors convey through a whole arsenal of nuanced emoting. Like Marx, Charlie Chaplin began his career as a stage actor who needed to broadcast emotion across the footlights. Other performers, from Theda Bara to Leonide Massine, employed the same device. Stylized exaggeration of facial expression is also the foundation of Kabuki and Noh makeup.

"The eyebrows form but a small part of the face," wrote the Greek orator Demetrius in the fourth century B.C.E., "and yet they can darken the whole of life by the scorn they express." The very word *supercilious* comes from the Latin word for the eyebrow and refers to the disdainful raising of this patch of hair to express contempt. In his autobiography Vladimir Nabokov uses this term to elegantly contrast the brows of two families: "The supercilious or surprised Nabokovs have rising eyebrows only proximally haired, thus fading toward the temples; the Korff eyebrow is more finely arched but likewise rather scanty." The arched brow conveys a great deal. T. H. White, author of *The Sword in the*

Stone, describes in his children's novel *Mistress Masham's Repose* a man who

> merely raised his eyebrows thus:
> ?

Even one brow can express a mood. In Rex Stout's series of Nero Wolfe mystery novels, narrator Archie Goodwin can lift a single skeptical eyebrow, which he does primarily to annoy his employer, who lacks the talent. Darwin referred to this ability when recording one of his many careful observations of instinctive expression in the people around him: "I noticed a young lady earnestly trying to recollect a painter's name, and she first looked to one corner of the ceiling and then to the opposite corner, arching the one eyebrow on that side; although, of course, there was nothing to be seen there." Darwin chronicled such observations over many years.

Our bestial brows, and creationist ideas about them, played a little-known part in inspiring Darwin to write *Expression*. Charles Bell, the pioneer English physiologist, had published during the first half of the nineteenth century three editions of his influential *Anatomy and Philosophy of Expression*. Bell believed that God gave humanity particular facial muscles to aid communication between his favorite creations, and Bell attempted to promote this belief to a hypothesis by marshaling evidence on its behalf. By the time Darwin wrote *Expression,* he strongly disagreed with—and took pains to refute—a number of Bell's ideas. Darwin was not opposed to anyone's religion, but beliefs presented as scientific evidence were open to rebuttal. (Even today this is the point at which scientists sigh in exasperation and reply yet again to creationist claims—when they are touted not as articles of faith but as interpretations of data.) In 1867 Darwin expressly stated to Alfred Russel Wallace his desire to "upset Sir C. Bell's view." The science historian and Darwin biographer Janet Browne explains the motivation behind Darwin's remark to Wallace: "If Darwin could demonstrate another, more practical purpose for such facial muscles, a purpose that clearly linked us to the rest of nature, he would weaken Bell's and all other natural theologians' arguments and strengthen the case for expression as a product of evolution."

Darwin rejected Bell's claims about voluntary muscular control and his assertion that humans have a wider range of expression than animals do. And there was another important point: The eyebrows are raised primarily by a muscle called the *epicranius frontalis*, pulled down by the *procerus*, and drawn together by the *corrugator supercilii*. Bell considered the corrugator "the most remarkable muscle of the human face. It knits the eyebrows with an energetic effort, which unaccountably, but irresistibly, conveys the idea of mind." On this theme, Darwin adds, Guillaume-Benjamin Duchenne de Boulogne described the corrugator as the muscle of reflection; Darwin himself considered it, because of its role in a number of expressions, "the muscle of difficulty." Bell also thought that the corrugator and some other facial muscles were unique to *Homo sapiens* and that together these distinct gifts endow us with a "combination of muscular actions of which animals are incapable." He was wrong. In his own copy of Bell's *Anatomy*, Darwin scribbled, "I suspect he never dissected monkey." As Darwin pointed out, our hairy kin also frown and scowl and flash their eyebrows—although other Victorian naturalists might have insisted that while exerting these same muscles, our fellow primates still look less intellectual than we do.

While recording his observations and his theories about our eyebrows, Darwin could not resist contrasting his own methods with some of the fanciful remarks about eyebrows made in the past. For example, he quotes the seventeenth-century painter Charles Le Brun on the expression of fright: "The eyebrow, which is lowered on one side and raised on the other, gives the impression that the raised part wants to attach itself to the brain, to protect it from the evil the soul perceives; and the side that is lowered, and which looks swollen, seems to be placed in that position by the phantoms that pour forth from the brain, as though to shield the soul and protect it from the evil it fears. . . ."

Darwin dimissed Le Brun's fancies as "surprising nonsense" and contrasted them with his own observation-based theorizing.

Yet eyebrows remained symbolic in Darwin's era. By the time of his death in 1882, the term *highbrow* had come to mean "intellectual" or "cultured," and soon it was also used to mock the pseudointellectual or culturally pretentious. Within a few decades the term spawned an antonym, *lowbrow*. There was one inevitable coinage left, and it came along after World War I. While sounding like a term from Tolkien, *middlebrow* designated any person or idea that supposed highbrows dismissed as conventional. Almost by definition, lowbrows were not applying such terms to themselves or to anyone else. The English humor magazine *Punch* cleverly defined middlebrow in 1925: "It consists of people who are hoping that some day they will get used to the stuff they ought to like." Tentative neologisms such as *broadbrow* and *mezzo-brow* quickly became extinct.

For Cesare Lombroso, a nineteenth-century Italian criminologist, one of the physical manifestations of a degenerate criminal body type was eyebrows that "tend to meet across the nose." Bram Stoker described his Dracula as if painting over a sketch by Lombroso, and even the vampire's brows fit the model: "His eyebrows were very massive, almost meeting over the nose, and with bushy hair that seemed to curl in its own profusion." This is not an uncommon occurrence. Frequently the eyebrows grow almost together, forming one long, dark hedge, a phenomenon celebrated most famously in the striking unibrow of the Mexican painter Frida Kahlo, which she memorialized (and exaggerated) in many self-portraits. Some ethnic groups have hairier eyebrows than do others, and a greater tendency to blur the space between them. At times, in ancient Greece and Rome, such brows were prized, and some women cursed with two distinct brows united them with kohl or some other cosmetic paint. Both Ovid and Petronius refer to women's fake eyebrows made of fur.

In many areas of the world, however, the response to the hair above the eyes has been an attempt to reduce it rather than expand it. There is an amusing exchange about women remaking their eyebrows in Shakespeare's late play *The Winter's Tale.* Mamillius is the young son of King Leontes of Sicilia. When his mother, Hermione, finds the youthfully energetic Mamillius bothersome and tries to foist him off on a lady-in-waiting, his reason for rejecting her and choosing another affords Shakespeare an opportunity for one of his irrelevant but observant asides.

MAMILLIUS: I love you better.
SECOND LADY: And why so, my lord?
MAMILLIUS: Not for because
    Your brows are blacker; yet black brows, they say,
    Become some women best; so that there be not
    Too much hair there, but in a semicircle,
    Or a half-moon made with a pen.
SECOND LADY: Who taught you this?
MAMILLIUS: I learn'd it out of women's faces.

Although it was a lovely bit of texture for Shakespeare to mention it, Mamillius did not have to be a genius to notice this common habit. Evidence of cosmetic attention to the eyebrows is as old as civilization itself. Supposedly Cleopatra revised her brows to suit her whims and the style of the period—a style that is visible in countless paintings and sculptures from the ancient Mediterranean. If we can trust the impressively realistic-looking Fayum mummy portraits, which were painted on panels over the wrapped bodies in the first century C.E., Roman women were elegantly attired but did not trim their noble brows. In the mid-1500s an Italian monk named Firenzuola held forth on the feminine eyebrow in his *Dialogo della Bellezza delle Donne* (Dialogue on the Beauty of Women): "The line of the brow should not be flat but curved in an arch toward the crown of the head, so gently that it is scarce to be perceived; but from this boss of the temples, it should descend more straightly." By the time of the Renaissance, women frequently shaved and plucked their eyebrows down to ethereal lines. Certain Madonnas of the period, such as Neroccio de' Landi's highly stylized *Madonna and Child with Saints John the Baptist and Catherine of Alexandria*, display the delicate calligraphic brows that would later arch over the faces of Jean Harlow and Marlene Dietrich. Some commentators observe that eyebrows are naturally less prominent on women, so that the plucking of them accents a feminine signal. However, as science writer Daniel McNeill points out, many whole cultures around the world also pluck their brows, from Amazonian tribes who disdain facial hair entirely to Hanafi Muslim women who pluck for religious reasons. Every visible aspect of the human body can become a new kind of signal by being either exaggerated or reduced or removed: culture playing with what nature gives it.

Such tinkering still thrives. Many women and men groom their eyebrows, shaping them in response to changing fashions. Sophia Loren's sultry beauty was enhanced by dark, rich brows in the 1950s, but a decade later she shaved them off and carefully drew them back on. In close-up her brows of this phase look like a drypoint sketch by Dürer. History's ebb and flow of eyebrow fashions is accelerated in novelty-hungry fashion and cinema. Throughout the history of film, actors and actresses have influenced the styles of the anonymous crowds who sat in the dark and admired their giant faces on a screen. Audrey Hepburn, Isabella Rossellini, Brooke Shields, Dorothy Dandridge—all have seen their distinctive eyebrows copied at large. For a while, in eyebrows as elsewhere, careful modification reigns supreme; then periodically high fashion admits back into the club more natural brows.

Most writing about the eyebrows focuses on those of women, but men are giving ever more care to their eyebrows. Think of the skeptically arched brow of a tuxedoed William Powell, mixing a martini as Nick Charles in *The Thin Man*. Clark Gable, Elvis Presley, Sean Connery—it is difficult to imagine their faces as being quite so impressive without the dramatic brows. So much attention to two small patches of hair justifies a question by the English poet Matthew Prior: "For what's an eye without a brow?"

THE BLINK OF AN EYE

While the eyebrows are jumping around to express each passing emotion, the eyes themselves are constantly opening and closing. It is a hardworking part of the face. The concentration of hair-trigger alarms in the eyelid further demonstrate the eye's importance in the anatomical hierarchy. Only one millimeter deep, even slightly translucent, the eyelid is the thinnest skin on the body; some scholars suggest that eye shadow was originally worn to disguise the veins visible through the eyelid. Yet this seemingly insubstantial layer of skin is packed with tools. Specialized nerve cells are densely clustered to defend against changes in temperature, the approach of foreign bodies, and the possibility of attack. In *The Expression of the Emotions in Man and Animals*, Darwin wrote about the eye's automatic self-protective gesture in response to sudden noise. "Starting . . . is accompanied by the blinking of the eyelids so as to protect the eyes, the most tender and sensitive organs of the body. . . ."

When not called upon for rapid security measures, the eyelid reliably performs the simple but crucial blink. A blink takes about a sixth of a second, and we blink roughly every four seconds. If we do not do this—if we try to resist the autonomic urge to blink—we find the eyeball drying out quickly. So, in order to keep the eye wet and functioning as smoothly as its ancestors did in their aquatic world, we momentarily blind ourselves. Blinking keeps the eye clear of dust and other particles that float in the air; the eyelid sweeps across the surface of the eye like a windshield wiper cleaning glass. To do this efficiently, the eye provides its own lu-bricant. Just as auto manufacturers learned to add glass-cleaning fluid that spurts onto a windshield to assist mechanical wipers, so does the eye provide a fluid that the eyelid can spread across the cornea. Without the utilitarian lubrication that gives them their romantic sparkle, our eyes might squeak when lidded like wipers on a dry windshield. In her *Sonnets from the Portuguese*, Elizabeth Barrett Browning expresses the liquid allure of living eyes when she says that the marble eyelids of death are not wet.

The lacrimal gland, which secretes tears, is located above and to the lateral side of the eye, away from the nose. It works constantly, producing tears that trickle down through minute tubules onto the surface of the eye. In its periodic blinks, the eyelid spreads the fluid evenly across the surface, then quickly rises and performs the job again a few seconds later. This liquid contains solutes, including proteins and antibacterial chemicals that help protect the surface. The tears washing the eye are gathered by two small ducts, one in the medial (inner) corner of each eye. These are visible. Many people think that these ducts, which you can see if you get close enough to the mirror, create the tears. Instead, with the tears washing down from above, the ducts actually serve as rainspouts to gather the accumulated moisture and channel it into the lacrimal sac in the side of the nose. From there it drains into the nasal cavity. About a quarter of the moisture produced evaporates from the surface of the eye, but the rest finds its way into these ducts or spills out onto the face in the form of tears. The resulting nasal drainage is why, when a sad movie or wind or an infection causes tears, we sniffle as the water accumulates.

In order to allow light into the back of the eye, where an image is created on the retina and from which the optic nerve carries the image to be analyzed and interpreted by the brain, the cornea—the outer layer of the front of the eye—must be transparent. Like all other parts of the body, however, the cornea requires oxygen. Most of the body gets its oxygen through blood, but blood is not transparent, and blood vessels feeding the cornea would inhibit vision. In its endless tinkering with the work in progress, evolution has adapted different kinds of eyes to different situations. Most mammals have evolved a singular adaptation, provided by the lacrimal gland, to meet the eyes' need for oxygen. It arrives dissolved in tears.

The human imagination has responded to even this tiny membrane that protects the eye. For one thing, the blink and wink have come to symbolize brevity. We complain that we did not sleep a wink (a phrase already in use by Cervantes and Shakespeare) and call the resulting afternoon nap "forty winks." The infinitesimal amount of time required to blink can represent an equally small amount of something else. In the final chapter of *Moby-Dick*, Ahab muses that the sea has not changed a wink since he first saw it as a boy. Because a blink represents the eye's time-out—a momentary blindness—the English scientist John Tyndall could thunder that "it is as fatal as it is cowardly to blink facts because they are not to our taste."

Sometimes we take ideas about blinking to fanciful extremes. A character in Fay Weldon's collection *Wicked Women* complains, "Therapists, New Agers and Born Again Christians seldom blink. A blink marks the mind's registration of a new idea. Converts have no intention of receiving new ideas." In some parts of rural Britain, a person who had the power to bestow the evil eye on another was called a blinker, and a sick or milk-poor cow was said to have been blinked. Apparently this kind of symbolic action of the eyes still works. Barbara Eden's character in the 1960s TV comedy *I Dream of Jeannie* crosses her arms and blinks to project her magical powers.

Noticeably frequent blinking can result from either physiological or psychological causes; a winking tic is a common manifestation of unease. A slower rate projects the opposite connotation—confidence, strength. The blink is an automatic action, but it responds instantly to the manual override of conscious decision: You can blink slowly or quickly merely by deciding to. We can bat our lashes to flirt or slow a wink to serve as a teasing signal or a gesture of camaraderie.

Few of us notice the inborn habit of blinking, either in ourselves or in others, until the eyelids' fleeting gesture of open and close varies its normal rhythm. In the 1990s American television host Bryant Gumbel complained that newsman Dan Rather made him nervous because Rather never blinked. Frequently this kind of unblinking stare is actually *intended* to make the opponent nervous; one dictionary defines the adjective *unblinking* as "showing no signs of emotion." Scientists have recorded staring contests among various primates, and anyone can observe them on a school playground. This tactic shows up especially when a literal or figurative staring contest emerges during a power struggle. In October 1962, during the Cuban Missile Crisis, U.S. Secretary of State Dean Rusk remarked with terrifying flippancy about the nuclear standoff with the Russians, "We're eyeball to eyeball, and I think the other fellow just blinked."

THE DREAMING EYE

> *While we are asleep in this world, we are awake in another one.*
>
> SALVADOR DALÍ

The speed of the blink prevents us from regarding it as a closed eye. Our diligent eyes seldom close for long except during sleep, and even then they keep moving. Surely many observant insomniacs over the centuries noticed that the eyes of a person dreaming sometimes move around behind the closed eyelids. Yet it took scientists centuries to realize the significance of this phenomenon. For a long time, both psychologists and physiologists dismissed sleep as purely unconscious downtime that required no further examination. Not until 1913 did the French psychologist and neuroscientist Henri Pieron publish the first serious examination of the physiology of sleep, *Le Problème Physiologique du Sommeil.* Soon the American physiologist Nathaniel Kleitman, considered the first scientist to focus entirely on our daily phase of unconsciousness, was spending the nights of the Roaring Twenties watching other people sleep. For this tiring work he earned the soubriquet "the father of American sleep research"— and, despite his sleepless nights, he lived to the age of 104. His pioneer studies of circadian rhythms included experiments with sleep deprivation and cycles of wakefulness during the night, as well as aspects of sleep in different populations.

This topic attracted some bright graduate students to Kleitman's laboratory in the Department of Physiology at the University of Chicago. In 1952 a student named Eugene Aserinsky was the first to monitor and attempt to decipher the eyes' periods of movement during sleep. He convinced Kleitman that the rapid needle movements on the oscillograph recorded actual eye motion rather than merely electrical artifact noise generated by the vacuum-tube amplifiers. Kleitman worried that their peers would be skeptical of these results. Aserinsky then proposed making a film of the movements, which were visible with the naked eye. Unfortunately the motion-picture equipment of the time was as bulky and inefficient as the medical monitoring equipment; often camera noise woke the subjects. Aserinsky's assistant, William Dement, did most of the camera work and eventually distilled the hours of nighttime film—less than 5 percent of which involved eye movement—into a three-minute documentary, the first visual record of these actions during sleep.

Such external evidence of a dream state gave researchers visible cues of when to interrupt dreams in various stages, so that they might quiz the dreamers about what their eyes were so energetically following. Aserinsky and Kleitman's advance was the beginning of serious dream research. Rapid eye movement (REM) during sleep proved so significant that other periods of unconsciousness came to be designated by its absence: NREM, non-REM. During the later 1950s, Dement explained what is now called the dream cycle and established the connection between dreaming and the REM part of the sleep cycle. The phenomenon invited a number of theories. Over the decades scientists reached a provisional consensus that attributed to REM a crucial role in learning processes and memory formation. Some researchers suggested that REM helped maintain *cortical homeostasis,* the ongoing maintenance of the brain's cerebral cortex. Theories of

REM's link to brain development grew out of the discovery that human infants experience an average of eight hours of REM daily, with this amount diminishing over time to an average of less than two hours daily for adults. In the early 1980s, Graeme Mitchison and Francis Crick proposed that REM might instead play a part in forgetting rather than remembering, by helping the brain clear its circuits of unneeded random associations. Recent evidence, however, seems to upset the long-standing notion that sleep is essential for memory formation.

When we think in narrow human terms, we formulate theories based upon our presumed uniqueness, and consequently we miss the deeper understanding provided by a more representative basis of comparison. Many other animals experience REM during sleep. In 1958 William Dement published the first paper documenting sleep cycles in creatures other than humans—cats. Most mammals experience REM cycles, including blind moles and near blind bats and deepsea whales. Some birds experience REM, but their cousins, the reptiles, do not seem to at all.

Formerly popular ideas about the function of REM in dreaming and memory were further challenged by the discovery in 1998 that the duck-billed platypus experiences REM more than any other creature—eight hours per day. The platypus is one of the most primitive known mammals—it even lays eggs—and its place at the top of the list contradicts all ideas that REM is a recently evolved function that participates in learning and memory in higher mammals. Extended periods of REM have been recorded in other creatures, including the thick-tailed opossum, giant armadillo, big brown bat, and golden hamster.

This accumulation of data suggests that, whatever REM does, it is not tied to the ability to learn responsive (rather than activate instinctive) behavior. Confirming this conclusion are the surprisingly short periods of REM experienced by the elephant and the giraffe. "The fact that the platypus displays REM is important for two reasons," writes Jerome Siegel, a UCLA neurobiologist and psychiatrist. "The first is that it seems to indicate that it uses sleep to allow its brain to develop/recharge itself. The second, which is possibly the more significant, is that REM has been around since a very early stage in animals' evolution." Now scientists are thinking that perhaps REM is involved in basic brain-stem functions.

Today sleep research is a booming field. In more than two hundred sleep-disorder centers around the country, scientists and physicians address everything from the effects of late-night shifts on worker safety to the consequences of sleeplessness in aging, cardiopulmonary function, and depression. Researchers date the genesis of this burgeoning field to the discovery of the nature of rapid eye movements during sleep. By serving as a window into our very dreams, the eye—messenger, guardian, metaphor—had served yet again as a window into, if not our souls, at least our ancient kinship with our fellow creatures. Dreams dance above the heads of many unconscious animals. While our sleeping eyes actually peer only at the inside of the eyelids, our dreaming eyes gaze at imaginary scenes like rapt moviegoers.

CLOSE YOUR EYES

> O death, come close mine eyes. . . .
>
> FROM AN ANONYMOUS
> SEVENTEENTH-CENTURY MADRIGAL

We close our eyes to blink. We close them to sleep. And, as the madrigal quoted above indicates, over the millennia enough people have closed their eyes as they died to inspire the recurring image of death as a kind of sleep.

In 1827 William Blake lay dying. He was almost seventy years old. A poet, engraver, painter, mystic, and certifiable nutcase, the critically neglected Blake had not yet been promoted to his current job as overworked visionary. And "visionary" is the word for him. Not only does much of Blake's poetry beautifully hosannah the visual splendor of the world, but all his life he actually experienced visions. He insisted that as a child he saw the prophet Ezekiel "under a green bough" and that he saw a tree "full of angels." He even called the highest level of creativity the Divine Vision. Therefore it is appropriate that the last anecdote we have about him involves his eyes. A moment after Blake died, his friend George Richmond walked up to the poet's staring face and gently closed the unseeing eyes. Richmond said he did it "to keep the vision in."

CHAPTER 4

Lend Me Your Ears

THE EAR OF A THOUSAND LI

There is a hierarchy among the facial features. The eyes and
mouth get all the glory. Like the nose, the ears are a body part
with which many people express dissatisfaction. The story is told
that, when Queen Elizabeth II viewed her newborn grandson
William for the first time, at Princess Diana's bedside in 1982,
she expressed the opinion of millions of observers when she said,
"Thank goodness he hasn't got ears like his father." If the queen
did not say it, she should have; Charles's pitcherlike ears are the
delight of cartoonists. Nor is he alone in this distinction. Clark
Gable actually went to the trouble of taping back his sizable ears
in some films, and still they were prominent enough to be paro-
died in a Looney Tunes cartoon.

In the late nineteenth century, the American architect Louis
Henry Sullivan uttered an immortal line: "Form ever follows function." Absorbed into the zeitgeist, this maxim is now chanted
everywhere from haberdashery to economics as the more alliter-
ative "Form follows function." Sullivan—designer of skyscrapers,
proponent of the aesthetic movement called functionalism—was
not referring to nature. His pronouncement, however, applies so
beautifully to evolution that it has long been co-opted by biologists. Evolution has curved our outer ears to gather sound waves

from the air and funnel them inside the head. The medium was biology, but the sculptor was physics. Gustav Eckstein said it with his usual offhand lyricism: "It took a long time to evolve an ear, many a hot afternoon of primal labor, the earth younger then." Eckstein was referring to the sense of hearing, but his remark applies equally well to the shape of the visible ear. This showy ornament on the human head is also called the *auricle,* from the Latin for ear, or the *pinna,* Latin for feather or wing, a term employed by biologists for everything from the winglike fins of walruses and seals (grouped under Pinnipedia) to the featherlike branching of pinnate leaves.

Nature had indeed been laboring a long time before it got to the showy ears of the chimpanzee and British royalty. The many shapes of ears in nature demonstrate the unique adaptations of our own. Always determined to be the most science-fictional creatures on the planet, insects wear their ears scattered all over their bodies. Moving up through the orders of animals, we find other variations older than our own ears. Amphibians and some reptiles are missing even the tunnel that leads from the eardrum to the outer surface of the head. Birds lack external ears; what look like ears on owls are actually mere tufts of feathers. Only mammals possess the seashell of cartilage and skin that constitutes the outer ear. Our comrades in this order certainly sport some stylish ones. To be impressed with the range of options available in the mammalian catalog, you need visualize only three: the jackrabbit, the African elephant, and the leaf-nosed bat. The elephant and jackrabbit have disproportionately large ears to facilitate heat loss in warm climates; bats generally have large ears to serve as better satellite dishes for receiving the echoes of their own sonar blips.

The shape and location of the ears of pachyderms and rodents evolved together, as did the shape and location of our own. Like our eyes, our ears demonstrate the body's tendency toward bilateral symmetry. This is elegant composition, but what is its purpose? The answer to this, too, comes from physics. As sound waves approach the head, inevitably they reach one ear before the other. Gathered by the outer ear, the differing sound waves pass down the ear canal to the middle and finally to the inner ear, where they are turned into electrical impulses. The differences in the way the two ears receive a sound enable the brain to triangulate its location. This method is the aural equivalent of the overlapping fields of vision in our two eyes, whose teamwork provides us greater depth perception. The satellite-dish effect works particularly well with sounds that fall within the range of human speech, indicating that, of all the clamor around us, we are factory-built to listen most to each other. The dual-receiver hearing system can detect differences of sound direction as small as 2 degrees. If you think of your head at the center of a 360-degree circle, you can imagine the fine-tuned hearing going on. You can hear a child walking barefoot across carpet and know from which direction she is approaching. You can hear a pin drop and, by the sound of its bounce, approach close enough to find it with your eyes. And finally there is also the matter of fail-safe redundancy. Like two eyes—and, for that matter, like our paired lungs and kidneys, hands and feet—dual ears provide insurance against mishaps in a busy world.

We demonstrate the curved-pinna principle every time we improve upon the ear's gathering abilities by turning toward a sound and cupping a hand behind the ear. It was also the rationale for the ear trumpet, in the days before electronic hearing aids. Human beings perform the gesture with the cupped hand all over the world. A telling evocation of this act appears in illustrations from Chinese mythology. The Empress of Heaven is assisted by two aides who, with their enhanced senses, see and hear the incidents occurring throughout the whole vast world (much the way that Odin, in Norse mythology, sends out his ravens Hugin and Munin—Thought and Memory—to perform the same reconnaissance survey). These assistants are named Eye of a Thousand Li and Ear of a Thousand Li. A *li* is a unit of measure that, although of varying length in different areas of China, equals roughly a third of a mile; obviously this gentleman is equipped with a sound-gathering array to rival the radiotelescopes at Arecibo. Just as Eye of a Thousand Li shades his eyes with his flat hand, so Ear of a Thousand Li usually is shown with his hand cupped around his ear.

Although we are touring the outside of the body, let us glance inside the cavern of the ear, for as far as we can see with the naked eye. The tunnel framed by the shell of the ear and visible for an inch or so inside the head is called the *external auditory meatus*. A meatus is a passage or canal; it is also the term for the opening of the urethra in the tip of the penis. Besides channeling sound waves into the middle ear, where they activate the devices that do the actual hearing, the meatus serves as an obstacle course. Foreign objects trying to enter the ear must evade both the sentry hairs near the opening and the waxy secretions of the ceruminous glands. Few invaders penetrate into the dark interior.

As a reminder of the significance that can lurk in humble themes, consider the grotty topic of earwax. The scientific term for this substance is *cerumen*, and there are two types. One is sticky, brown, and wet; the other, brittle and grayish or beige. In general, Africans and Europeans possess the former and Asians the latter. These distinctions are more important than you might guess. In the late 1990s, attentive scientists at the Nagasaki University School of Medicine noticed an odd correlation: Japanese breast-cancer rates are higher in women whose ceruminous glands produce sticky earwax. This observation suggests that the earwax gene, or other genes connected with it in some way, influences the development of the disease. After all, the ceruminous glands are physiologically related to the apocrine glands that produce milk in the breasts. With swabs and notebooks in hand, the scientists went to work. They have found that wet wax is a dominant trait and dry recessive, but at the time of this writing the correlative gene still eludes researchers.

These outstanding accessories have assumed many guises in religious and artistic symbolism. In the beginning was the word, and there had to be an ear to hear it; so the ear became one of the receptacles of the breath of life. Egyptian mythology had the left ear receiving the "air of death" and the right ear the "air of life." In classical mythology snakes lick the ears of Cassandra, Hellenus, and Melampus, instilling in them the ability to foretell the future. Numerous symbols join forces in Christian iconography when the ear serves as a doorway into the soul of the Virgin for a dove signifying the Holy Ghost. In some cases a ray of light from heaven enters Mary's ear, by which aural sex she conceives the Logos, the Word.

Chaldean diviners invented a whole framework of portents based upon abnormal births, and many of their signs concerned the ear. A small right ear or a wound below the right ear warned that the house of the father would be destroyed; both ears missing predicted mourning and a diminished country; and if both ears were merely deformed, it meant that the country would fall and its enemies rejoice. An infant born with the Picassoesque condition of two ears on the same side of the head augured a calm reign and a prosperous nation. The rarity of these sports of nature may explain why calm and prosperity always seem so fleeting.

Even freed of mythological baggage, by virtue of its bodily importance the ear still gets to appear in many of the great scenes of literature. In *Paradise Lost* the angel whispers metaphysics into Adam's ear, just as the serpent hisses rebellion into Eve's. His father's ghost informs Hamlet that the incestuous, adulterate Claudius poured a toxic substance into the porches of his ears while he was sleeping. The dead king has a quick vocabulary: *porch* derives from the Latin word for a passage or entrance, and the ear has represented the gateway to the brain, the heart, and even the soul.

Thomas Carlyle said that he was persuaded to stop smoking by a "long hairy-eared jackass" of a physician. Why do hairy ears seem uncouth? They are a common sight. As men lose the hairs on their heads, the foyer of the ear provides a site for new ones to sprout as a further affront to dignity. Hairs grow inside the ear throughout our lives. In later life, like hardy weeds, they expand their range to the very porch of the meatus and even climb the curves of the pinna itself. Indeed, an often overlooked part of the ear is named for some of these hairs. The little lip of cartilage at the front of the ear, which seems like a half-open door guarding the entrance, is called the *tragus*. It is not unique to human beings. In many bats, such as the long-eared and the leaf-nosed and the pallid, the tragus is greatly enlarged. Apparently its increased size assists in the reception of sound waves coming from the side rather than from the direction in which the ear is pointed. It is the door of the tragus that we push in to seal our ears against unwanted advice and the enticements of sirens. The tragus's name derives from the hairs behind it; the Greek *tragos* meant literally "he-goat."

Like most other body parts, the ears have been measured and analyzed as possible clues to character. Anthropometry, the measurement of the proportions of the human body, did not disregard the ears. Alphonse Bertillon, a criminologist in nineteenth-century Paris, attempted to incorporate ear measurements into his system of anthropometric identification. Like our fingerprints—and like more obscure signature patterns, including facial heat emission—our ear patterns are unique. Unfortunately, they are not distinctive in ways that are as easy to sort and file as are fingerprints, and they are also less cooperative in providing prints of their contours. The curves of the ears do fall into several categories, but no correlation between those categories and behavior has ever been established. Yet apparently this idea survives. In John Irvin's 1995 film *A Month by the Lake*, Vanessa Redgrave's character repeatedly tells an attractive man that she judges people by their ears. She is attempting to deduce character from generalizations about shape.

Although neglected in porn films, the ear is an erogenous zone for many human beings. Seldom do we allow others to touch our ears, and sexual foreplay is one of those occasions. For some people a whisper in the ear can be as erotic as a kiss on the lips. Cultural history has not ignored the ears' power to inflame. Once upon a time, the female ear was considered a near obscene imitation of the female genitals, its coils slyly imitating the labia, its channel dangerously reminiscent of a different orifice. The idea also materializes in ancient mythology, in which birth from the ear is an occasional theme. This method was how Karma, offspring of the Hindu sun god Surya, entered the world. "Grotesque" is a mild term for this concept, until you realize that birth from a seashell was also a common theme. Spiraling, folded inward, both ear and shell seemed visually reminiscent of the vulva—and therefore symbolic of both sexual intercourse and birth. As a result, through sympathetic magic, seashells have sometimes been considered charms to aid in delivery of a child. In a reversal of this imagery, nineteenth-century slang dubbed the vulva "the ear between the legs."

Whether created in the image of God or not, ears had to be covered to mask their vulgar allusions and shield our innocent youth from corruption. This prohibition applied only to female ears, although male ears were identically shaped. Why the creator of the universe would bother to place visual-pun erotic sculptures on the sides of the head was a topic seldom addressed. Perhaps it was the same artistic flourish, an indulgence of divine whim, that resulted in some human beings' sporting little points on their ears.

DARWIN'S POINTS AND HOUDINI'S WIGGLE

In his 1871 masterpiece *The Descent of Man*, Charles Darwin credited the sculptor Thomas Woolner with bringing to his attention what Darwin called "one little peculiarity in the external ear." While sculpting Puck with pointed ears, Woolner realized that many human ears have a small rounded point projecting from the upper edge of the helix, the inward-folded margin—the curved outer lip—of the external ear. Woolner examined the ears of various human beings and then the ears of some of our simian cousins. In time he communicated his discoveries to Darwin.

As the great biologist described them in the *Descent,*

> These points not only project inwards towards the centre of the ear, but often a little outwards from its plane, so as to be visible when the head is viewed from directly in front or behind. They are variable in size, and somewhat in position, standing either a little higher or lower; and they sometimes occur on one ear and not on the other.

It was not difficult to find evidence of this peculiarity among apes and even among the more distantly related monkeys. Darwin theorized that the helix was the folded-in former margin of the ear. Next he speculated that the folding was in some way related to our external ears' being, for whatever reason, flattened against the sides of our head. He cited incidents of seemingly atavistic ears whose outer edge was not curved inward to form a lip but instead flattened and pointed like that of a quadruped. He concluded, "If, in these two cases, the margin had been folded inwards in the normal manner, an inward projection must have been formed." Diligently quoting the objection of a colleague that these results were "mere variability," Darwin nonetheless maintained his own view that "the points are vestiges of the tips of formerly erect and pointed ears." To this day the protuberance on the edge of the helix is called "Darwin's point" or "Darwin's tubercle." It is now known to be a dominant trait that is inherited in a simple Mendelian manner. Unlike, say, height, the shape of Darwin's point is not influenced by environment but determined entirely by the combination of genetic factors.

It is interesting to follow the progression of Darwin's description of the primitive ear. Five years after the *Descent,* the Italian criminologist Cesare Lombroso published the first edition of his *L'Uomo Delinquente* (Criminal Man). Lombroso argued that most criminal behavior is atavistic, a reversion to evolutionarily primitive actions. This early manifestation of social Darwinism was founded in the same sort of thinking on heredity that led an

American authority on prisons to pontificate, "Good seed gener-
ates sound and healthy fruit, and imperfect parentage can only
yield defective offspring." Lombroso based his theories largely
upon unsubstantiated data and his own limited group of subjects,
which consisted mainly of Sicilian prisoners. He argued that
criminals in general are born, not made—and, like most of us, he
was adept at seeing what he wanted to see. For example, Lombroso flatly declared that criminal personalities have larger jaws
and cheekbones than do the virtuous masses. Admitting that this
trait might not be atavistic, Lombroso speculated that it could result from "the setting of the teeth or tension of the muscles of
the mouth, which accompany violent muscular efforts and are
natural to men who form energetic or violent resolves and medi-
tate plans of revenge." He casually threw around statistics that at
best we might describe as questionable: "Prognathism, the projection of the lower portion of the face beyond the forehead, is
found in 45.7% of criminals."

Lombroso insisted that ears, like so much of the rest of the
body, exhibit outward clues to degenerate personalities. In this
contention, too, his evidence was less than overwhelming. He
said that in criminals the ear was "often" larger but "occasionally" smaller. "Twenty-eight per cent of criminals," he recited,
"have handle-shaped ears standing out from the face as in the
chimpanzee"; and then he added, with no visible embarrassment,
"in other cases they are placed at different levels." Confident of
the generalities resulting from his biased data, Lombroso continued his parade of statistics that nowadays strike us as patently
self-refuting. All along, maligning this supposed atavism, he was
playing into the fear of our animal nature exemplified throughout mythology, in which one of the bestial attributes of satyrs is
their pointed ears.

Eventually Lombroso got around to Darwin's point in the
Descent:

> Frequently too, we find misshapen, flattened ears,
> devoid of helix, tragus, and anti-tragus, and with a

protuberance on the upper part of the posterior margin (Darwin's tubercule), a relic of the pointed ear
characteristic of apes. Anomalies are also found in the
lobe, which in some cases adheres too closely to the
face, or is of huge size as in the ancient Egyptians; in
other cases, the lobe is entirely absent, or is atrophied
till the ear assumes a form like that common to apes.

The Lombrosan worldview thrived in fiction, its native habitat. In 1897 Bram Stoker employed the entire catalog of degenerate characteristics when he described his very embodiment
of evil, Dracula. Along with teeth that protrude over his lips, a
lofty forehead, and eyebrows that almost meet over his nose, the
count's "ears were pale, and at the tops extremely pointed."

Human ears differ in another interesting way from those of other
creatures. Rabbits, dogs, horses—they can move their ears, turning them automatically in the direction of sounds. This talent
does more than amuse their fellows; it is essential in tracking the
sound waves produced by approaching predators or fleeing prey.
Even nocturnal prosimians flaunt this enviable trick. However,
like other diurnal primates with smallish sound gatherers pinned
to the sides of their heads, we have lost the ability to turn our
ears. A few people can wiggle theirs, but seldom impressively.

In *The Expression of the Emotions in Man and Animals*, Darwin
cited many examples of the ways that other creatures move their
ears. Horses, boars, even rabbits pull back their ears prior to at-
tacking, yet sheep, cattle, and goats do not. Darwin even distinguished between a horse that was turning its ears toward a sound
and one laying its ears back in anger. He then went on to examine
the various ways that animals raise their ears and turn them toward
sounds. "The head being raised," he added, "with erected ears and
eyes directed forward, gives an unmistakable expression of close
attention. . . ." Because we have other options, human beings
manage attentive expressions without drafting our ears to help.

Darwin addressed the human loss of ear movement in another work, *The Descent of Man*. He pointed out that the extrinsic muscles that move the external ear are rudimentary and vary in the degree of their development: "I have seen one man who could draw the whole ear forwards; other men can draw it upwards; another who could draw it backwards; and . . . it is probable that most of us, by often touching our ears, and thus directing our attention towards them, could recover some power of movement by repeated trials." As far as utility is concerned, this trait has all but vanished in *Homo sapiens*. Acknowledging the usefulness of mobile ears to animals, Darwin argued that in human beings the "whole external shell may be considered a rudiment, together with the various folds and prominences (helix and antihelix, tragus and anti-tragus, &c.) which in the lower animals strengthen and support the ear when erect, without adding much to its weight."

Other scientists, as Darwin himself pointed out, disagreed. Some, for example, suspected that the outer ear's cartilage transmitted sound waves to the "acoustic nerve." What Darwin could not solve was the question of why our cousins, as well as our progenitors, had lost the ability to raise or turn their ears. It is not a loss that we go around lamenting, because our flexible necks allow us to turn our entire head toward incoming sound waves. In the long run, this method actually may be more efficient.

Aware of the mobile ears of our fellow creatures, to this day we say of ourselves that an interesting comment makes us prick up our ears. Yet two millennia ago Pliny the Elder was stating flatly in his *Historia Naturalis*, "Only man has ears that do not move and this is the origin of the nickname 'flap-eared.'" Actually, Pliny used the term *flaccus*, a word that translates as "hanging down." He also quoted his contemporary and fellow encyclopedist, Pompeius Trogus, who thought that humans exhibited several external signs of character, including this helpful observation: "Large ears are a sign of one who talks too much and is silly." Even the usually gullible Pliny is forced to mutter in a stage whisper worthy of Groucho Marx, "So much for Trogus."

Now and then someone possesses an atavistic ear-wiggling talent. In 1898, in his first known publication in a magazine about conjuring, a twenty-four-year-old Ehrich Weiss—already officially known as Harry Houdini—explained various subtle methods he employed to communicate to his assistant whatever information a member of the audience was writing on a pad or slate. "I have even," he boasted in the article, "trained my right ear to move up and down to thus give my assistant the tip."

NOR THE EAR FILLED WITH HEARING

Some parts of the body started out with one purpose and in time moved on to another, as if they had been hired for a certain job and later promoted or reassigned. Just as etymology demonstrates that history lies fossilized in our every utterance, so does the body preserve a record of the past in its structure and behavior. "Evolution never starts from a clean drawing board," says the English biologist Richard Dawkins. "It has to start from what is already there." In doing so it preserves aspects of each organ's previous function.

The ears are a perfect example of this phenomenon. Because our hearing apparatus evolved from organs that had a previous role, it was not free to develop in just any old way. Inside the skull, between the satellite dishes on each side of the head, lie the real hearing devices, which double as organs maintaining equilibrium. Guarded by the strongest bones we possess, deep within the innermost spirals of the cochlea, the inner ear carries a drop of the primordial sea in which our distant aquatic ancestors evolved. Like its antecedent organ, the ear responds to pressure changes in the surrounding fluids. However, now those fluids are contained and are responding to subtle vibrations from sound waves. Our ears are so brilliantly attuned to hearing that the organ of Corti is insulated against the faint sound that blood makes when moving through capillaries.

Astronauts on the moon and instruments on Mars express a consensus about those places: They are as silent as the grave. On the moon we played like children, but only in mime, because the airless atmosphere refused to carry our laughter. In contrast Earth is a noisy planet. For inconceivable millennia before we pulled ourselves upright and began to flirt and lie, water rushed, birds sang, wind howled, trees fell, and snow crunched under the feet of animals. At this instant there are millions of living beings out there making noises at each other and at us. Guided by these sounds, we have hunted prey and dodged predators and slowly evolved our complex cultures.

Nowadays, armed with our flexible, durable, comical, seashell-shaped protuberances, we still face toward the invisible sound waves that the world aims at us every day. Seemingly without effort, the brain processes these signals into meaningful communication. It happens everywhere, constantly, to every human being who is not deaf. Thanks to the atmosphere's ability to carry sound waves, creatures evolved mechanisms with which to perceive them, and we are allowed to rejoice in Sidney Bechet's clarinet and Annie Sellick's voice. Henry Thoreau once noted in his journal the surprising way that we are moved by sounds: "A slight sound at evening lifts me up by the ears, and makes life seem inexpressibly serene and grand."

The most impressive talent of the ear is its ability to distinguish between a surprisingly wide range of sounds, stretching across 130 decibels. It is easy to throw around this kind of statistic, but we ought to take a moment and examine it. The term *bel*, which is used also in measurements of voltage and power, measures changes in the intensity of a sound; normally, however, we use a decibel (dB), a tenth of a bel. Like the Richter scale for tracking earthquake intensity, the measurement of decibels is logarithmic. With the lower threshold of human hearing labeled 0, an increase in sound of 10 dB would be ten times as loud, but 20 dB would be ten times ten or a hundred times as loud. A logarithmic scale spanning 130 decibels means that the human ear at its best can accommodate a 10 trillionfold difference in loudness.

Of course, this kind of range has no point unless we can re-member the distinctions. Scientists estimate that on average the brain can distinguish among four hundred thousand sounds on file in the wet gray database between the ears. We recognize an immense array of sounds produced by inanimate and animate na-ture, not to mention the many awful and beautiful sounds that we make on our own. The Roman poet Virgil reminds us of a more urgent message unspoken behind the noises hurrying daily into our ancient pinnae: "Death twitches my ear. 'Live,' he says; 'I am coming.'" Until death twitches our ears, we can enjoy an aspect of hearing described in the Book of Ecclesiastes. Follow-ing the famous verse "All the rivers run into the sea; yet the sea is not full," the Preacher continues this line of thought: "The eye is not satisfied with seeing, nor the ear filled with hearing." Just as our hardworking eyes never overflow with the world's insistent visibility, so do our glorious ears never fill up with sound.

CHAPTER 5

A Ridiculous Organ

> It might appear prudent, if not altogether necessary, to
> commence by vindicating the Nose from the charge of being
> too ridiculous an organ to be seriously discoursed upon.
> But this ridiculousness is mere prejudice; intrinsically one
> part of the face is as worthy as another. . . .
>
> GEORGE JABET, *Notes on Noses*, 1852

As Jabet realized, the nose gets even less respect than the ears do.
And yet, as vehicle of memory, hidden alarm system, secret arbi-
trator in seduction, and silent partner in taste, it harbors within
its cartilaginous precincts a number of fascinating talents. The
nose is also a multitalented figure of speech. It can be out of joint
or poked in where it doesn't belong; you can lead someone by it
or pay through it; and apparently certain masochists hold theirs
to a grindstone. We say that we turn up our noses at things we
dislike and sniff out bargains and smell a rat. Obvious proposi-
tions are as plain as the nose on your face. Some Americans
claim that if your nose is itching, it means that company is coming. According to Hippocrates, a nosebleed, like the onset of
menstruation, foretells the breaking of a fever.

Our overactive brain stays busy, always making up stuff about
everything around it, but the *real* nose is no less versatile than its
cultural stand-ins. Over the centuries it has tracked our game

and warned of putrefaction in carrion. Nowadays we still rely upon it to detect everything from smoke to gas leaks. Our lives would be immeasurably depleted without the nose's ability to discern the volatile molecules of the world around us—"volatile" because, when you smell something, you are identifying the particles that are already floating away from the object. For this reason odor can be an informal guide to biological instability: the scent of pollen versus that of glass, excrement versus stainless steel. We consciously use this sense less than do most other creatures, although perfumers and vintners rely upon it. Many animals employ it—insects, for example, and fish and mammals. Odor is one of the most primitive forms of communication. A female silkworm moth can advertise her amorous moods for several miles, and in response a male can home in as if locked on to a radar beacon. A whole different sensory map of the world hovers in the minds of even the cats and dogs around us, brought to them by their quivering nostrils.

We take our own nostrils for granted, but they are interesting developments. Nietzsche declared that his genius resided in his nostrils, and Proust might well have said the same about himself. Nietzsche meant his sense of smell, but the physical nostrils are interesting in themselves. The word *nostril* comes from the Old English *nosthyrel,* "nose hole." These paired orifices are handily situated for the analysis of airborne molecules and outfitted with many hairs to act as filters against larger particles and the occasional misguided insect. The nostrils are charmingly mobile and often attractive. Just think what they would be like if we were descended from reptiles instead of mammals. We would either hibernate or migrate and would remain always as cold-blooded as a Republican legislator, and our nose holes would not wiggle. Pliny the Elder quoted Megasthenes' claim that among the Nomads of India there was a people called the Sciritai who, "like snakes," had only holes in place of actual nostrils. Imagine our nostrils stilled. Without their sensitive, expressive nostrils, deodorized American tourists could exhibit no quiver of revulsion at the proximity of sweat on foreign subways.

"And the Lord God formed man of the dust of the ground," intones Genesis, "and breathed into his nostrils the breath of life; and man became a living soul." In our own time, we have demoted the nostrils from the role of divine conduit, as indicated when humorist Calvin Trillin mocked the 1960s films of Andy Warhol and Yoko Ono by claiming that he was planning a documentary to be entitled *Nostril.* In the 1960s television comedy *Bewitched,* Elizabeth Montgomery's character could not enact her mischievous witchcraft without a twitch of her nose. Her signature gesture was a cue for the viewer that narrative conventions were about to be suspended because magic was afoot, but it was possible because of the human nose's similarity to those of our fellow creatures.

Our mammalian heritage bequeaths the nostrils a lively animation, but for many people the nose just perches in the middle of the face, holding up eyeglasses and harboring congestion. Despite the magical nose twitch of TV witches, human beings in general lack the wildly mobile noses of our hairy kin. Dogs, horses, cats—these animals are fun to watch partially because their busy noses contribute so much to facial expression. The external movement symbolizes the internal busyness. Any animal demonstration of powerful smelling reminds us of how little our own noses can do by comparison. In a bit of doggerel that sounds as if it might have inspired the patter song in *Singin' in the Rain,* G. K. Chesterton permits a dog to pity our relatively deodorized lives:

> *Even the smell of roses*
> *Is not what they supposes. . . .*
> *And goodness only knowses*
> *The Noselessness of Man.*

Even Aristotle was already complaining that we have lost our sense of smell, but we seem noseless only in comparison with the profound sniffing of our fellow creatures. For decades scientists have been slowly working out the development of our olfactory system. It begins early. Roughly five weeks after conception, a nasal pit forms in the proto-face of a human embryo, and, within a couple of weeks, distinct nostrils appear. Olfactory neurons form early on, but they don't develop into specialized cells that respond to odors until about twenty-eight weeks after conception. At this point the fetus actually has a functional sense of smell; scent flourishes even in the womb. Premature infants born before this age exhibit no response to odor, but at about this time they begin to draw toward or away from scents. "Olfactory abilities rapidly improve during the third trimester, and festuses' olfactory life is surprisingly rich," writes neurobiologist Lise Eliot. "Their sense of smell is not impeded by amniotic fluid, since odor molecules normally enter a liquid phase—the nasal mucus—before binding to their olfactory receptors."

These nasal passages remain damp throughout life, and frequently they become clogged by allergies or infection. Sometimes the resulting substance is dry and sometimes wet, and both must be removed from the nose. For the dry substance, from *Seinfeld* we learn that undoubtedly Moses was a nose picker because of the dry desert air of the Holy Land. For the wet version we have coined, from the Old English *gesnot*, a repulsive four-letter word. This substance must be removed, or gravity will remove it, which is why it contributes to the charming personal hygiene of small children. "Life is made up of sobs, sniffles, and smiles, with sniffles predominating," says O. Henry. Many sniffles are caused not by sadness but by congestion. The ancient Greek physician Galen thought that the substance emerging from the nose represented "a purging of the brain." The Greek word *katarrhous,* meaning "to stream down," evolved by the Middle Ages into *catarrh,* an all-purpose term for respiratory ailments such as allergies and sinus troubles, any infection of the nasal membranes that would cause discharge.

What to do with the substance that accumulates in your nostrils? In *The History of Manners,* the cultural historian Norbert Elias explores the evolution of attitudes toward the natural functions of the body during the last half millennium or so. Over the centuries Europeans and their far-flung descendants became less and less comfortable with revealing the body and exhibiting any of its natural processes or by-products. Recorded strictures range from a casual 1589 reminder not to urinate against indoor walls to a strict 1774 injunction that it is "never proper to speak of the parts of the body that should always be hidden." One of the behaviors that Elias documents is nose blowing. "It is unseemly to blow your nose into the tablecloth," wrote a fifteenth-century German author, and at about the same time a French commentator instructed, "Do not blow your nose with the same hand that you use to hold the meat." Among his many writings about childhood and civility, the sixteenth-century Dutch humanist Erasmus intones the following advice: "to blow your nose on your hat or clothing is rustic . . . nor is it much more polite to use your hand. . . . It is proper to wipe the nostrils with a handkerchief, and to do this while turning away, *if more honorable people are present.*" The class-conscious emphasis is his own. By the end of the eighteenth century, an author called Le Mésangère was writing in *Le Voyageur de Paris,* "Some years ago people made an art of blowing the nose. One imitated the sound of the trumpet, another the screech of a cat. Perfection lay in making neither too much noise nor too little." Over time we seem to have forgotten how to play this particular wind instrument.

One of the more amusing aspects of the epidemic UFO silliness of the last few decades is the resemblance between spacefaring visitors and the dominant mammal on Earth. With a completely different evolutionary history, from a galaxy far away, somehow aliens still have two eyes and a mouth and a head on a neck and two hands and two feet—and even, usually, a nose. In their famous story from the 1960s, Betty and Barney Hill disagreed in their separate accounts of the aliens who forced momentary notoriety upon them. Barney said that in place of noses the visitors had merely slits. Apparently the extraterrestrial trauma confused one of them, for Betty insisted that the aliens had *huge* noses, even specifically mentioned that they had noses like Jimmy Durante's.

Speaking of fiction, the nose shows up as a quirky old character actor in a lot of it. Perhaps the best-known nasal image in art is to be found in the honesty meter that causes so much trouble for the wooden puppet, and entertains so many Freudians, in Carlo Collodi's horrific *Pinocchio*. (The Disney studio toned down Collodi's story for the animated version.) Nikolai Gogol, ever exercising his admirably free imagination, tells the story of the barber Yakovlevich, who one morning finds—what a disgusting image—a nose in a loaf of fresh bread. Meanwhile, on the other side of town, one Kovalev looks in a mirror and finds that "instead of a nose he had a perfectly smooth place!" In the dystopian future society of Woody Allen's early film *Sleeper*, the Leader has been killed and must be cloned from all that remains of him—his nose. The science-fiction writer Damon Knight, in his story "God's Nose," has the Creator sneeze our universe into existence. In *Smeller Martin*, a children's book by Robert Lawson, the author of *Rabbit Hill*, a young boy gets into all sorts of adventures because his nose can smell so many things that other people's cannot. And, according to legend, in tenth-century England the yard was defined as the distance from the tip of King Edgar's nose along his outstretched arm to the end of his middle finger. Similarly, the inch is said to have begun as the length of the knuckle on the royal thumb.

Whether Edgar's nose really contributed to history seems uncertain. Other incidents are less questionable. As we usually think of him, the romantic but grotesquely schnozzed Cyrano—cinematically incarnated by actors from Jose Ferrer to Steve Martin—is fictional, but Edmond Rostand and other writers drew their inspiration from the adventures of the real M. de Bergerac. Rostand places in the mouth of Cyrano the following apotheosis of the nose:

> *A great nose indicates a great man—*
> *Genial, courteous, intellectual,*
> *Virile, courageous.*

The Japanese would have disagreed with this sentiment, or at least would have redefined "great." In 1853 Commodore Matthew

Perry wielded the threat of U.S. attack to extort trade agreements
from Japan, and the next year a Japanese artist caricatured him
with a huge nose—the Western facial feature that many Asians
found amusingly oversize.

In England the very next year, George Jabet, whose apologia
for the worthy nose provides the epigraph for this chapter, published a book entitled simply *Notes on Noses*. "We believe that
besides being an ornament to the face, a breathing apparatus, or
a convenient handle by which to grasp an impudent fellow," declared Jabet of the nose, "it is an important index to its owner's
character." Three-quarters of a century before, a Swiss mystic
and theologian named Johann Kaspar Lavater had founded the
movement called physiognomy, the reading of personality and
character from the shape of facial features. Jabet found his inspiration in Lavater. But he had his own taxonomic bent, and he indulged it in classifying the nose into categories. One genus was
nationalistic and racist: "Every nation has a characteristic Nose;
and the less advanced the nation is in the civilization, the more
general and perceptible is the characteristic form. . . . [T]he most
highly organized and intellectual races possess the highest forms
of Noses, and those which are more barbarous and uncivilized
possess Noses proportionately Snub and depressed, approaching
the form of the snouts of lower animals, which seldom or ever
project beyond the jaws."

Jabet enumerated five different species of European nose. Each
had different ramifications on men and women. The Roman or
Aquiline on men reveals decisiveness and energy but also a short-
age of refinement, and on women it imparts an unseemly "masculine energy." On both men and women, the Greek or Straight nose
is naturally "the highest and most beautiful form," which denotes
a refined and artistic personality. In women, Jabet volunteered, it
might reveal itself in artistically composed needlework. The self-
evident Cogitative, which is sometimes classifiable only from the
front, appears more on men than on women because women live
by their emotions rather than in thoughtful meditation. The Jewish or Hawk nose, which on men inevitably betokens shrewdness

both in worldly affairs and in profiting therefrom, also seldom appears on women, who are naturally dependent and credulous. Two final categories, the Snub and Celestial noses, were counted together. On men the former indicates pettiness, cheek, and a generally weedy character, and the latter is a variation only slightly leavened with a vulpine cunning. Such weakness of character in women "is excusable and rather loveable," and Jabet generously confesses "a lurking *penchant*, a sort of sneaking affection which we cannot resist, for the latter of these in a woman."

The "science" of physiognomy almost ended the scientific career of the young Charles Darwin before it began. The incident took place more than two decades before Jabet's catalog of informative noses. If they did not regard him as the devil incarnate, biblical literalists might see Darwin's life as divinely nudged toward his accomplishments. Few historical figures seem to have been so destined for their fate. He failed at university courses in law and in medicine. Turning toward the safety net that caught many a gentleman's son, he decided to settle down as a country parson and cultivate natural history as a hobby. Then, in 1831, at the tender age of twenty-two, he was suddenly offered the unpaid position of captain's companion aboard the survey ship HMS *Beagle*—but only after other people had turned down the opportunity. Later Darwin learned that even when he interviewed with the captain of the *Beagle*, the harsh and puritanical Robert FitzRoy, he barely overcame another hurdle. FitzRoy, wrote Darwin in his *Autobiography*, "was an ardent disciple of Lavater, and was convinced that he could judge a man's character by the outline of his features; and he doubted whether anyone with my nose could possess sufficient energy and determination for the voyage." Darwin's personality and letters of recommendation overcame FitzRoy's fears, and in time the captain thought so highly of the adventurous young naturalist that he even named geographical features after him. "I think," added Darwin, "he was afterwards well-satisfied that my nose had spoken falsely."

CHAPTER 6

The Archaic Smile

LOUIS ARMSTRONG'S LIPS

The sixth and fifth centuries B.C.E. were busy around the Mediterranean. Heraclitus was trying to unite fire and misanthropy and flux into a worldview. Pythagoras was formulating his elegant hypotenuse theorem and creating the vocabulary of scientists' much later quest for a mathematical Theory of Everything. Phidias was coaxing the human shape from marble, Callimachus supposedly inventing the Corinthian column. Pindar and Sophocles were burning the midnight oil to scribble their visions of sacred games and royal incest. And there were all those nonwriters harvesting food and raising children and fighting wars.

Despite so much activity, we know relatively little about the era, because time has pilfered most of the evidence. Aeschylus, for example, is thought to have written sixty-odd plays; only seven survive. Countless mysteries remain. "The enigma is symbolized to perfection," scholar Peter Green points out, "by that famous if baffling phenomenon, the Archaic Smile." You can see this curious artistic trend in several different examples of the *kouros*, a popular ancient mode of sculpture portraying the bust or entire body of a young man. These figures look cartoonishly primitive beside the elegant proportions of later sculptures that we think of as "classically" Greek. In contrast to the more recent works' solemnity, the earlier figures also look disconcertingly *happy*. Bearded or beardless, with various hairstyles, they have one thing in common: Their lips curve upward in what is almost a smirk. Scholars judge that this artistic convention originated in Anatolia, the peninsular region of Turkey also called Asia Minor, but by the middle of the sixth century B.C.E. it was common throughout the region "—only to disappear again," notes Green,

> rather like the Cheshire Cat's grin, during the age that followed. No one really knows what the Archaic Smile was all about, though it has confidently been held to symbolize anything from man's divine spark to a happy extroversion supposedly typical of the Lyric Age.

Whatever inspired this convention, it is charming to think of Aegean sailors spending their off hours smilingly posing for a sculptor. However, those upturning lips playing their forceful role in human expression are far older than the word *archaic* seems to convey. They are older than *ancient*, perhaps even deserving *primordial*. The smile, and the mouth's many other movements, goes much further back into prehistory than any ancestor that even vaguely resembles *Homo sapiens*. The quick nervous grin of a scampering macaque, the broad smile of a chimpanzee performing a practical joke, a human infant's first reflexive smile at her mother—these gestures remember an ancient lineage.

Lips are useful. The lower jaw is the only bony part of the head that moves: the cartoon mandible chopping up and down, as in the talking skulls of movies. Our mandible does not open to the extremes of some creatures'; it cannot disengage and permit the passage of larger prey, as it does in certain snakes. All it does is open and close and grind sideways a bit. Yet we are satisfied with the jaw's limited repertoire, because our mouths—especially our lips—can do all sorts of fancy things.

The evolution of our clan's hairless faces seems to have followed the development of a trait that has always helped us to see monkeys and apes as disconcertingly similar to ourselves—extremely mobile lips. The hard-times admonition to "keep a stiff upper lip" means to prevent it from quivering, which in turn reassures those around you that you are not about to cry or otherwise admit to emotion that some may consider inappropriate. Because expressions are fleeting and defined by movement, we can limit ours by deliberately resisting the automatic motions of our facial features, especially the eyes, brows, and lips.

Most mammals, as a glance at your dog or cat or guinea pig will confirm, lack these curiously flexible lips. They can bare their teeth, but our wider range of expression is not available to them because the upper lip is more closely attached to the gum. The limitations of lip movement among animals have always been a stumbling block for filmmakers wishing to portray non-human characters as talking in a humanlike manner. Think of the awkwardly speaking horse in the television series *Mister Ed.* Even though the barnyard creatures in the film *Babe* are masterfully done, the animals who most convincingly speak are the young chimpanzee couple and the elderly orangutan in the sequel, *Babe: Pig in the City.* Their mobile upper lips, genealogically similar to ours, nicely shape human words.

During the surge of animal-communication experiments over the last few decades, this shared trait inspired scientists to manipulate apes' lips to pronounce human syllables. Unfortunately, the animals proved less than enthusiastic about the idea. In one famous experiment, scientists molded the lips of a chimpanzee named Viki until she could add an "m" sound to the "ah" she already produced, resulting in her pronunciation of the word *mama.* At first Viki required the Pavlovian signal of a human finger on her lips to make her speak the word, but in time she began to recite it at random. Apparently she had no hint of its meaning.

Viki was sensitive to the touch of a finger on her lips in the same way that we are. Touching requires something to be touched;

for this reason many stories about body parts merge. Nursing, for example, is no more about the breasts than it is about the mouth. Babies who are born with a blocked esophagus must be fed artificially, and sucking on a rubber teat proves to calm them as much as nursing at a breast calms other babies. "It had to be a case of contact for contact's sake," remarks Desmond Morris. "So the touching of something soft with the mouth is an important and primary intimacy in its own right." The body begins preparing for this ability early. By the eighth week after conception, the milestone at which the embryo gets promoted to a fetus, the future human being is still only an inch long but already has doll-like, recognizable features—toy arms and legs, of course, but also eyes, nose, and even miniature lips.

The outer edge of the lips is an important borderland, the line of demarcation between the skin of the face and the entirely different mucous membrane that lines the mouth and alimentary canal (which is similar to the mucous membrane lining the vagina). The lips are one of the more sensitive parts of the body. Their importance in its cooperative system of government is emphasized by the numbers and kinds of touch receptors they house. The skin of the lips has roughly twenty times as many receptors as, say, the legs. Our skin worries more about cold, which can fatally reduce body temperature, than it does about heat, which our internal systems can combat with perspiration and respiration. To warn against heat loss, clever structures called *arteriovenous anastomoses* are clustered in crucial areas, especially in the eyelids, nose, lips, hands, and feet. This network, however, provides only distress signals, not security guards. You can see the result on your lips. After the alarms inform it, the brain instructs other employees to shut down the perimeter and thereby restrict heat loss. The anastomoses are a system of connections between veins and arteries. In response to changes in external temperature, they bypass capillaries and alter blood flow to accommodate the needs of a given area. Although their subcutaneous work is invisible, you can observe its result on the surface. If you are cold enough, the rerouting of the red blood will leave your lips and fingernails bluish.

Lips come in many sizes and shapes among ethnic groups and among individuals within each group. They have long been a focal point of cosmetic anxiety, and over the last few decades a number of African Americans have sought lip surgery to diminish the size or shape of their lips. Historically, African Americans have undergone less cosmetic surgery, for several reasons—dark skin's tendency to form keloid scars, surgery's inability to disguise racial origin (unlike the way that "corrected" epicanthic folds can strikingly diminish the appearance of Asian ancestry), and the relative poverty of a largely disenfranchised population. And yet, as Elizabeth Haiken explains in *Venus Envy*, her history of cosmetic surgery, even lip surgery can be motivated by a yearning for assimilation: "Race- and ethnicity-based surgery has always focused on the most identifiable, and most caricatured, features: for Jews, noses; for Asians, eyes; for African Americans, noses and lips." This part of assimilation is one of the interesting aspects of Michael Jackson's complete restructuring during the 1980s and '90s. "The results of his surgery," remarks Haiken, "suggest that if he is not trying to look white, he is at least trying to look less black. . . ." Whatever its tangle of motivations, this sad flight from his own ethnicity in a star of Jackson's magnitude transmitted a pulsar-size signal to African American youth—at least until his star faded and he merged with the rest of the pop constellation.

Occasionally over the decades, Caucasians have undergone lip-reduction surgery merely to banish the *suspicion* that they might be tainted with what one doctor in the 1920s called "the question as to Ethiopian origin." One white woman was convinced, apparently solely upon the basis of her "flat" nose, that she was the result of a black man's rape of her mother. Some surgeons resisted these patients' entreaties and advised psychological counseling. Other surgeons experienced no such qualms. In 1946 one wrote about his work on correcting such "congenitally unesthetic" lips.

Sometimes lips, like so many other body parts, simply wind up as fashion victims. Like the earlobe, the soft lip seems to invite a deliberate wounding. Many body parts, from eyebrows to labia, were pierced with rings as part of the 1990s trend that has been called our version of millennial self-flagellation. Yet this sort of lip piercing is not as innovative as contemporary pundits seem to think. In the late eighteenth century, when the French explorer Jean-François de la Pérouse visited British Columbia, he found Tlingit women wearing what he called *labrets*—lip plugs. These huge plugs of bone and ivory, or even of stone and wood, could extend the lower lip outward from the face as much as a few inches. If you try it yourself, you will discover that our lips and cheeks can stretch an impressive distance.

Tlingit females began receiving lip modifications shortly after birth. Sometimes as early as the age of three months, a girl's lower lip was slit and the wound blocked from healing. Increasingly large ornaments were inserted; the sag of the lip grew ever more extreme, until it interfered with eating, drinking, and speaking. When the French tried to abolish the custom, they reluctantly admitted that, from their point of view the empty sagging lips were no great improvement over the plugged version. But the custom had significance in its original culture. The plug size denoted status, and only free women could wear it.

During the next century, Charles Darwin remarked in smug Victorian mode, "As the face with us is chiefly admired for its beauty, so with savages it is the chief seat of mutilation." Then he blithely went on to quote the way that "savages" considered enhancement of the face to improve beauty itself. New Zealand missionaries reported to Darwin the comments of some young native women regarding tattooing: "We must just have a few lines on our lips; else when we grow old we shall be so very ugly." Darwin told a further anecdote about how our easily modified lips could assist gender definition and status. The women of the Makalolo in southern Africa wore a *pelelé*, a large ring composed of bamboo and metal, through a hole punched in the upper lip.

David Livingstone, the Scottish missionary and explorer famously "found" by Henry Stanley, reported to the British Association that the inserted metal caused the upper lip to extend
sometimes as much as two inches beyond the tip of the nose. A
smile lifted the plug back over their eyes. Livingstone asked a
chief, Chinsurdi, why the women wore such ornaments. A busy
man who had little time for nonsense, Chinsurdi snapped, "For
beauty! They are the only beautiful things women have; men
have beards, women have none. What kind of a person would
she be without the *pelelé*? She would not be a woman at all with a
mouth like a man, but no beard."

Thanks to their role in both food processing and communication, lips have evolved several features that make them highly
versatile. This arrangement has helped produce some great art.
In the chapter on the skin we met Marsyas, inventor of the double flute, who is bested in a musical competition and subsequently flayed alive by Apollo. Like fables about Gilgamesh or
Jesus or the U.S. Founding Fathers, Greek myths appear in several versions. According to another, it is not Marsyas but the
goddess Athena who invents the double flute—which, as its
name indicates, is not to be confused with the multiple reeds of
the syrinx, the pipes of Pan (the instrument that much later lent
its name to the vocal apparatus of birds). However, when Athena
sees her reflection and realizes how distorted her face becomes as
she plays the instrument, she abandons her budding musical career. "The sound was pleasing," she admits in Ovid's account;
"but in the water that reflected my face I saw my virgin cheeks
puffed up. I value not the art so high; farewell my flute." In another story Athena's ever petty colleagues and rivals, Aphrodite
and Hera, mock her horn-blowing expression until she discards
the new instrument. Whether vanity or peer pressure triggers her
decision, at the moment when she chooses appearance over artistic expression, Athena fumbles her chance to join a more noble
assembly than the fractious country club on Mount Olympus.

Because they are made of sterner stuff than the trust-fund babies of the Greek pantheon, mortal performers not only ignore vanity but also overlook pain. They physically suffer for their art. Ballet dancers torture themselves to achieve the unnatural ability to run on the tips of their toes. Guitarists, cellists, bassists— all practice until their fingers bleed. Horn players of every sort complain of damage to their lips; the better the performer, the likelier that hours are spent in practice daily, and the greater the abuse to the lips.

Louis Armstrong was probably the best and most influential jazz trumpet and cornet player of the twentieth century, and the pain he suffered from his damaged lips is legendary. Our sensitive, flexible lips did not evolve armor to protect them from such torment. Photographs of Armstrong's later years not only record the battered condition of his lips but even catch him applying salve. By the time of a 1932 New Year's Eve concert in Baltimore, Armstrong's raw and swollen lips were already causing the thirty-three-year-old musician constant pain. Band members reported that he kept checking his lips in the mirror while applying salve. Saxophonist Milton "Mezz" Mezzrow later remembered watching Armstrong picking at the sores with a needle. "I been doin' this for a long time, you know," explained Armstrong. "Got to get them little pieces of dead skin out, 'cause they plug up my mouthpiece."

Mezzrow described Armstrong's performance that night: "He started to blow his chorus, tearing his heart out, and the tones that came vibrating out of those poor agonized lips of his sounded like a weary soul plodding down the lonesome road, the weight of the world's woe on his bent shoulders, crying for relief to all his people. . . . Everybody knew that each time that horn touched his lips, it was like a red-hot poker to him." When the audience's applause drowned out his last note, Armstrong stood there, holding his horn and panting, with his lips oozing blood that he licked away. By November of the next year, matters were even worse. Armstrong was performing at the Holborn Empire in London one evening when his lip split open, causing a steady drip of blood down the front of his tuxedo shirt. Finally he took a four-month rest. In 1934, with more lip troubles, he had to cancel so many concerts that his manager sued him for breach of contract. Ed Penney, formerly a prominent jazz disc jockey in Boston, remembers that he once asked Louis Armstrong his advice for trumpeters who hoped to follow in his footsteps. And the great Satchmo replied simply, "You got to keep up the chops."

Armstrong suffered from lip troubles for the rest of his life. You can hear a later result on *Ella and Louis*, the collaboration recorded in August 1956 for Verve. It features a toned-down Armstrong, sans both clowning and virtuoso trumpeting. Part of his restraint was in response to the dignified collaboration with fellow immortal Ella Fitzgerald, but part was the result of having to fall back on his voice. His lips hurt him too much to permit the kind of energetic horn playing for which he had become famous.

THE BADGE OF A HARLOT

> *Her lips were red, and one was thin,*
> *Compared with that was next her chin,*
> *Some bee had stung it newly.*

> JOHN SUCKLING

Suckling, a seventeenth-century English poet, was neither the first nor the last man to admire the swollen lower lip of a young woman. In *Lolita* the obsessive Humbert Humbert coos of his barely pubescent beloved that she has "lips as red as licked red candy, the lower one prettily plump." This very trait is a sign of Lolita's blossoming sexuality. Frequently it is tricky to differentiate the gender of babies and young children—which is one motivation for immediately dressing girls and boys differently, to begin the lifelong propaganda barrage about gender signifiers. When sexual hormones kick in at puberty, however, the differences become more apparent. For one thing, many girls' lips acquire a ripe, swollen look.

Neither bee-stung nor flushed-looking lips are limited to

pubescent females, but the combination occurs among them more than among any other subset of humanity, and most prominently when other messages are also being broadcast. Any gender-specific trait that blossoms only with the onset of sexual maturity must be examined for possible evolutionary significance as a mating signal. Apparently these newly reshaped lips are yet another visual cue of fertility, along with breasts and increased fat deposits around the hips. Also, as they become sexually excited, girls' lips, like their nipples, swell and become a brighter red. Up to this point, most scientists agree. Some biologists, however, theorize that the perpetually larger, redder lips of the maturing female mimic the reddening and swelling of their genitals during sexual arousal. Other scientists dismiss this notion as Freudian obsession disguised as rampant adaptationism. Yet nature is fully as outrageous as our most fevered imaginings, and quite as fixated on sex as the gentleman from Vienna. Some of our primate cousins *do* have features on their faces that mimic their aroused genitals. Large baboonlike monkeys called mandrills are famous for their brightly colored faces, an unusual trait in a mammal. They have grooved blue cheeks and clownish red noses. This Mardi Gras mask of a face is an uncanny imitation of the mandrill's own genitals.

The baboon's face, the product of millions of years of evolution, demands that we ask a question: What the hell is the point of having a face that mimics your genitals? The theory is that it allows your shameless genes to purchase further advertising space on the most prominent billboard available to primates: our attention-getting naked faces. As far as humanity has ever been able to determine, nature has no particular interest in us or our ideas of fairness. Like a huge corporation, it worries only about the bottom line. After all, what do the fit survive *for*? Nature gets right to the point with its personal ads. As soon as human beings are biologically capable of reproducing, they exhibit prominent flags of their sexy new status. The lips of adolescent girls unite with their new breasts, like the suddenly deep voices and hairy chins of their boyfriends, to proclaim, "I am ready to reproduce."

Unquestionably, the lips remain throughout life a major site for sexual advertising, much of it aimed at promoting youthfulness—and, by implication, fertility. Women of all ages emphasize their lips with attention-grabbing lipstick, usually from somewhere in the red end of the spectrum. Perhaps this form of ornamentation is yet another way that the human imagination, nature's great spin doctor, has taken a natural phenomenon and repackaged it in cultural mythology. Toward lips as with breasts, many men behave like sticklebacks and other creatures known to respond enthusiastically to secondary sexual characteristics that have been exaggerated beyond anything nature had thought to devise. "Lipstick," admits Debbie Then, a social psychologist, "has more of a sexual connotation than other makeup." Inexpensive and portable, it has long been the most popular form of makeup in Western society—indeed, the one item that many women consider essential. Jean Ford Danielson, the cofounder of BeneFit cosmetics, declared that the entire industry had grown up around lipstick. According to various reports, the average American woman (if there is such a person) paints her lips with several pounds of pigment in the course of her life. As Meg Cohen Ragas and Karen Kozlowski write in their cultural history of lipstick, *Read My Lips*, "Even women who don't wear makeup wear lipstick."

During the Depression the cosmetics industry promoted makeup as illusion, artifice to enhance appearance without being visible itself. Advertisements from the time record the straight-faced touting of a brand of lipstick that "can't make you look painted" because "it isn't paint." Some advertisers admitted that lipstick was a sexual come-on by encouraging its responsible use in the cause of snaring a lifetime mate. At first this trend toward artificial naturalism discouraged the application of lipstick in public—apparently equating it with a magician's explaining a trick—but by 1933 *Vogue* placed its seal of approval on applying lipstick by declaring it "one of the gestures of the twentieth century."

Although pigment stored in and applied with a retractable

metal tube was invented as late as 1915, women had been painting their lips long before the twentieth century. Ancient Chinese and Egyptian illustrations portray the act of coloring the
lips. Greek women twenty-five hundred years ago enhanced their
natural lip color with vegetable dyes, and Cleopatra famously
painted hers with carmine and henna. Of course, not everyone
has approved of women's attracting attention to their lips. In the
seventeenth century, England suffered from the sermonizing of a
hellfire-and-brimstone pastor named Thomas Hall. Like so many
others of his profession throughout history, Hall was terrified of
female sexuality and how it could lead poor, weak-willed men
into the perfumed arms of Satan. He denounced lipstick as "the
badge of a harlot" and insisted that women who painted their
lips were out to "kindle a fire and flame of lust in the hearts of
those who cast their eyes upon them."

When some women in the 1960s and '70s, impatient with ossified definitions of femininity, rejected provocative lipstick colors, companies responded by inventing more subtle shades. They
sold well and still do. There have been white lipsticks and even
Goth black. Obviously, not *all* wearing of lipstick can be traced
to the Circean tool-kit. Affecting some of its wearers as much as
it does their audience, lipstick can provide yet another pleasure
that would have frightened poor Thomas Hall. Véronique Vienne, author of *The Art of Doing Nothing,* wrote a lyrical apotheosis of lipstick sensuality:

> I may never have known the pleasure of wrapping my
> lips around substantial thoughts if it hadn't been for
> the buttery, vanilla-scented kiss of a glorious Guerlain
> lipstick. . . . The lush sensation of the lubricant on
> our mouths encourages us to weigh our words and ar-
> ticulate them carefully. As we enunciate, the scrump-
> tious balm stretches and constricts, stimulating tiny
> muscles that radiate from the lips to our jaws, cheeks,
> nose, eyes, and ears.

This luxuriating response sounds rather more pleasurable than the chaste Victorian habit of borrowing color for the lips by wetly kissing rose-colored crepe paper. Lipstick is clearly here to stay, with a cachet all its own. In a kind of visual synecdoche, lips sometimes represent the entire woman. One of Andy Warhol's tributes to Marilyn Monroe consisted of rows and rows of only her lipsticked mouth, and as long ago as 1934 Man Ray portrayed a pair of disembodied lips floating in the sky over a landscape. Lipstick is important enough symbolically to figure in a number of contemporary slang terms. Consider "lipstick lesbian," for example, a term used by both homosexuals and heterosexuals to describe lesbians who embrace some of the traditionally heterosexual signals of femininity in our culture. The expression has been extended to other groups not famous for their glamour, and now we have "lipstick librarian" and "lipstick feminist."

Sometimes lipstick does more than emphasize the lips and inflame their natural color. For generations many women have been artfully using lipstick to alter the apparent shape of their lips, sometimes accentuating the redrawn outline with lip liner. Actresses from Joan Crawford to Clara Bow popularized such reshaped outlines as the Cupid's bow and the hunter's bow. There was a resurgence in the use of liner during the 1990s, when actresses and singers exaggerated the youthful, pouty contours of their lips.

Color and outlining are not the only ways to enhance your sultry moue. Over the years zealous advocates have invented numerous methods to alter the size and shape of the lips. In the late nineteenth century, women who aspired to the fashionable ideal were taught to practice saying words that began with the letter *p*. This notion was disseminated by people such as Mrs. General, the vacuous widow in Dickens's *Little Dorrit*. She prescribes a whole mouth-puckering vocabulary: "Papa, potatoes, poultry, prunes, and prism, are all very good words for the lips: especially prunes and prism." Understandably, many women found this advice ludicrous. The great feminist activist Elizabeth Cady Stanton declared her impatience with women who wasted time honing their "prunes and prisms expressions."

Nor would Stanton have applauded the contributions of Baronne d'Orchamps, who in 1907 published *Tous les Secrets de la Femme*. Her suggested regimen for the mouth sounds tiring and also makes lips seem detachable: "Soak the lips for at least five minutes in a glass of warm water. Dry them, then smear them with camphorated pomade. After a quarter of an hour, dry them again with a soft cloth and put on some glycerine. . . . A light sucking of the lips, a little bite, will give to them in an instant a crimson bloom." In the years just before the First World War, the modish young women called Gibson girls sucked on hot cinnamon drops and bit their lips to improve their hue.

All of these faddish antics predated the age of cosmetic surgery, which permits us to even more drastically revise the image in the mirror. Nowadays both the painted illusion of fuller, rounder lips—as well as the acquired posture from reciting *p* words—can be surpassed with genuinely redesigned lips. There are specific cosmetic surgeries such as lip lifts and implants of silicone, collagen, or fat. So many film stars (and hopeful starlets) have undergone such surgery that the natural presence of prominent lips on an actress must be defended against accusations of self-sculpture.

Advertising attacks the self-esteem of potential customers. It convinces us that our shortcomings in appearance, wardrobe, transportation, and scent can be redeemed only through a certain product, which a company is selflessly producing to fulfill this already desperate need. Lipstick ads are no exception. One of many shameless advertising lines took the idea of cosmetically enhancing the lips' color and turned it into a requirement. In 1927 the new lipstick Rouge Baiser was promoted as a less moist form of mouth paint that was "designed to make you kissable."

A KISS IS JUST A KISS

*A kiss can be a comma, a question mark or an exclamation
point.*

MISTINGUETT,
legendary Folies Bergère dancer and singer

Contrary to the claims of Rouge Baiser, it does not take lipstick
to make you kissable. Over the millennia human beings have
kissed each other before mammoth hunts and after a hard day
building pyramids. The evidence indicates that if we could look
far enough back in time, we would find our hairy ancestors
smooching on the African savannah. Among hidden-away tribes
in various parts of the world, the verb for "kiss" is the same as the
verb for "smell," and scent is always an important part of a kiss.
In some countries friends and even strangers greet by kissing on
the cheek or on the lips. In ancient Persia men kissed their social
inferiors on the cheek and their peers on the lips. There's a great
line in Gilbert and Sullivan's *Ruddigore:*

> *I'll wager in their joy they kissed each other's cheek*
> *Which is what them furriners do.*

Some kiss the air near the cheeks. Lipsticked kisses have decorated the posters of teen idols from James Dean to the Back-
street Boys. People kiss their pets. The mambo singer Lou Bega
describes a day so fine that he has to kiss it. In some places it is
still considered a romantic gesture to kiss a woman's hand upon
meeting her. Bystanders laugh when Raskolnikov kisses the
ground at the crossroads to atone for his sin, but Pope John Paul
II has kissed the ground in many countries. Parents kiss their
children as they carry them off to bed. In Joyce's *Portrait of the
Artist as a Young Man*, Stephen Daedalus is still young enough to
notice such behavior: "His mother put her lips on his cheek;
her lips were soft and they wetted his cheek; and they made a tiny

little noise: kiss. Why did people do that with their two faces?" It is a question that occurs to many young children. It is also a question that they soon stop asking, because, as anthropologists have noted the world over, most cultures kiss in one way or another.

The prehistoric origins of such behavior can be seen in another papal rite: genuflecting supplicants kissing the pope's ring. Primatologists such as Jane Goodall and Biruté Galdikas testify to the prevalence of this very gesture (sans jewelry) among our closer relatives. Goodall, observing chimpanzees at Gombe, Tanzania, frequently witnessed lower-ranking males submissively crouch and kiss some part of the alpha male's body. In her classic account *In the Shadow of Man*, Goodall described a variation of the gesture: "I saw one female, newly arrived in a group, hurry up to a big male and hold her hand toward him. Almost regally he reached out, clasped her hand in his, drew it toward him, and kissed it with his lips." Galdikas, celebrated for her work with the orangutans of Borneo, describes a tender maternal version of this behavior. After a mother orangutan and her son cuddled, the young male began to walk away. A moment later the mother softly called to him, and the child returned. His mother placed her left hand on his head and extended her right in front of his face, and her son gently kissed it.

Darwin died long before these and other primatologists began to reap the fruits of their painstaking field studies. However, even as early as *The Expression of the Emotions in Man and Animals*, he reported an incident observed in a zoo: separately captured chimpanzees meeting for the first time and greeting with hugs and kisses. Darwin also quotes Richard Steele's remark about kissing: "Nature was its author, and it began with the first courtship." Apparently this epigram is not accurate even if you move the courtship back a few million years prior to what Steele had in mind. The prevalence of kissing in our primate cousins reminds us to look for its origins further back in time than mere cultural explanations would suggest. Kissing began before the first protohuman or primate courtship, and its comforting pleasure led to its reclassification as a romantic gesture.

Various primatologists have observed kissing as a routine

form of communication between chimpanzees. One field-worker watched a mother chimp reassure her worried infant with a humanlike maternal gesture—gathering him into her arms and kissing his head. Goodall observed a male unmercifully attack a female; then, when the female hurried after him and crouched in submission, he patted her head reassuringly and even leaned forward and kissed her brow. A prodigal son returned to the family fold and greeted his mother with an offhand brush of his lips on the side of her face. Although most chimp greetings are less demonstrative, longtime friends will hug and may kiss each other's neck or face. In short, kissing is a gesture of communication that even predates our definition of ancient. The British primatologist Carole Jahme sums up both its legacy and its technique: "Kissing and the sexual act predate language. They do not require speech to be successful; in fact, talking is superfluous."

The Romans distinguished between three categories of kisses: *oscula*, *basia*, and *suavia*, which translate respectively as "friendship," "love," and "passion." (Their judicial system also tried and punished the overly demonstrative for *crimen osculationis*—the perpetration of unwanted kisses—but only among citizens of equal social position.) A kiss as the culmination of a wedding ceremony dates all the way back to the Roman era. It was already the kind of kiss that Romeo bestowed upon the dead lips of Juliet, "to seal the immortal contract." As taxonomically precise as the Romans, the Talmud distinguishes between kisses that demonstrate leavetaking, respect, and greeting.

In the Old Testament, kisses are a common way to greet or bid farewell to relatives, although close friends might also kiss on such occasions. But kisses were not to be trusted even before a certain betrayal in the garden of Gethsemane. For example, Joab inquires after Amasa's health, grasps him by the beard to kiss him, and runs him through with a sword. The ultimate literary symbol of hypocrisy occurs in the New Testament. Jesus is alternately communing with his dad and berating his sleepy disciples when Judas enters the garden and identifies him for the priests and elders by kissing him.

The kiss does not fare well in the Bible. There are few references to erotic kissing in the whole anthology. One in the Second Book of Samuel is the story of a bad, bad woman who tempts a man from the straight and narrow by kissing him and alluding to her perfumed bed. "With the flattering of her lips she forced him," claims the author, invoking the age-old whine that Adam tried with God in the Garden of Eden: It's not *my* fault; *she* started it. Elsewhere we find a different attitude toward kissing, and for that matter toward temptation itself. It appears in one of the more controversial books of the Bible, the Song of Solomon or the Song of Songs. How this steamy collection of love poetry sneaked into the Bible is a mystery; biblical scholars call it a *hapax legomenon,* a term for a word or form that occurs only once. The Song of Songs casually ignores sin, God, and Israel. For many such reasons, it is unlikely that the author was actually Solomon; because he is mentioned, however, traditionally the work is attributed to him.

Uncomfortable with the openly erotic imagery, some Jews and Christians have unconvincingly argued that the whole love affair is an allegorical presentation of either God's love for Israel or Christ's love for his believers. To most people, though, the poems simply read like one of the purer expressions of delight in the body anywhere and one of the first in ancient literature. Amid the hormone-fueled mutual admiration that makes the poems sexy—busy hands, oral sex, lying all night betwixt breasts—are some memorable references to kissing. The first verse launches enthusiastically into the subject: "Let him kiss me with the kisses of his mouth: for thy love is better than wine." Later there is another unchaste reference to edibility: "Thy lips, O my spouse, drop as the honeycomb: honey and milk are under thy tongue." He praises her body with fruitful imagery, and she invites him into the orchard. It is this sort of lewd repartee that inspired the criticism directed at St. Teresa of Ávila, the sixteenth-century Spanish mystic: "It is not decent for a woman to write about the Song of Songs." This opinion prevailed even though at least one of the narrators is a woman.

❧

The Irish poet Louis MacNeice described all of London as being littered with remembered kisses. Indeed the whole world is littered with kisses remembered or anticipated. This gesture is mysteriously compounded of affection and desire. We daydream about kissing someone we yearn for. We steal kisses at parties, describe a letter as "sealed with a kiss" or scrawl XOXO on it, blow kisses out the windows of trains, kiss photos of loved ones. Sometimes we take this fondness in odd directions. The word *kissogram* dates back to the first half of the twentieth century and meant a valentine actually kissed by the sender; the term reemerged in the 1980s for messages delivered by provocatively dressed women and including an actual kiss.

It is not enough even to trace this habit back to our primate kissing cousins and ancestors. If we did not invent it, why did *they* start doing it? Some anthropologists and psychologists propose that kissing evolved from mouth-to-mouth feeding between mothers and infants. Sigmund Freud claimed that an infant's pleasure in suckling at the maternal breast is reactivated upon discovery of similar activities such as thumb-sucking and kissing. For Freud this marked the beginning of the oral stage of development, supposedly followed by the anal phase. Whatever percentage of Freud's frontier sketch of the psyche may correspond to actual landmarks, some such responses would have been developing long before our innovative primate forebears reached the point of naming themselves human beings. Inevitably, the insertion of the male tongue into the female mouth is seen as symbolic phallic penetration. For some people kissing is *more* intimate and affectionate than sexual intercourse. James Jones expresses this idea in *From Here to Eternity:* "The tabu said you never kissed a whore. They didn't like it. Their kiss was private, like most women's bodies."

For the puckering action that draws the lips together into a kissing position, we must thank the *orbicularis oris*, a sphincter muscle like the *orbicularis oculi* that surrounds the eye. It is a ringlike band of muscle that encircles the lips, extending down toward the chin and running between the nose and the upper lip. Some people call it the kissing muscle. With this muscular contraction we do interesting things. Even among romantic or sexually charged kisses, the variety is impressive. We can perform the barest brush against lips, no less intense for its brevity or light touch; slow romantic kisses with mouths closed or lips only barely parted; first-time reconnaissance missions that may predict the rest of the evening; marathon bouts that include tongue battles, nibbles, and the rest of the array exemplifying the urge to consume the object of desire. Many people have commented upon this craving. When Salvador Dalí met Elena Diakonova, who called herself Gala, he was a sexually innocent twenty-five and she a decade older and married to the poet Paul Eluard. They met at Cadaqués on the rocky coast of Spain. Years later Dalí claimed that before Gala he had never embraced anyone, and he sighed that their first kiss on the shore began "the hunger that drove us on to devour each other down to the very last morsel." It is amusing to find Charles Darwin, of all people, writing in a similarly personal tone about this hungry sort of kiss. His notes are preserved in the fragmentary volume that scholars call the N notebook, one of the offshoots that grew out of his early thinking on evolution. He scrawled in his usual quick way a few remarks about "ones [sic] tendency to kiss, & almost bite, that which one sexually loves." Darwin also notes that in both dogs and humans "sexual desire makes saliva to flow."

This hunger, which Freud dubbed *oral aggression*, is what scares people. It would seem that any activity that serves equally well as a prelude to reproduction, a form of greeting, and an evening's entertainment bears nature's highest seal of approval. Yet it is the kiss's very role as appetizer for a feast of sex that inspires fear. During the Spanish Civil War, the puritanical dictator Francisco Franco joined forces with the church to create a hypocritically moralistic regime that even prohibited public kissing. Hollywood's sex-obssessed Hays Office, guardian of cinematic morals during the Depression and afterward, limited the duration of kisses in movies for fear that osculatory excess would lustfully inflame American youth. Alfred Hitchcock circumvented this restriction in *Notorious* by instructing Cary Grant and Ingrid Bergman to periodically stop touching lips for a moment.

If you fear the dangers of sexual intercourse, then you are smart to fear kissing. In his 1971 book *Intimate Behaviour*, the zoologist Desmond Morris, author of *The Naked Ape*, examined the physical and social aspects of human mating. In doing so he schematized our typical animal courtship sequence into twelve steps. The dozen highlights graph the progress of increasingly intimate communication that most creatures follow, from visual to physical. They range from the first glance to the first touch, from an arm around the waist to a meeting of genitals. The first kiss falls in the middle of the scale. (In American high-school parlance, a kiss used to count as "first base," with actual intercourse considered to score a home run.) Morris articulates a consequence of kissing that every teenager knows from experience:

> Kissing on the mouth, combined with the full embrace, is a major step forward. For the first time there is a strong chance of physiological arousal, if the action is prolonged or repeated. The female may experience genital secretions and the male's penis may start to become erect.

Yes, that would be why there are so many restrictions against kissing. The parts of the body most luxuriously equipped with nerve endings include the lips, clitoris, and penis. Apparently they are networked like computer systems to facilitate communication between each other.

This common knowledge motivates the censure, but not every manifestation of fear has been political. In 1901 the Women's Christian Temperance Union launched a national campaign to educate the masses about the dangers of kissing on the lips. Not even invoking a scientific rationale in the name of puritanism could save such a doomed crusade, but Anna Hatfield, the union's energetic leader, nonetheless denounced the ancient habit of lip kissing as "barbaric and unhealthy." If the unhygienic savages out there simply could not refrain from slobbering all over each other, the least that they could do was disinfect their mouths beforehand.

Sadly, kissing does have its risks. It is certainly a good way to share a cold. Mononucleosis is called "the kissing disease" because kissing is also an easy way to transmit this highly communicable nuisance. Mono is caused by the Epstein-Barr virus, which is also implicated in chronic fatigue syndrome. Fear of it became one of the bogeyman warnings of the 1950s, in an attempt to curb teenage kissing. Soon, though, researchers discovered that most people in the world test positive for the virus, and the younger they are when they contract it, the less they suffer. Incidentally, the kissing disease is not to be confused with the kissing *bug*, a literal true bug, a bedbug, a bloodthirsty member of the assassin bug family that likes to bite human beings around the lips.

Mononucleosis can be communicated by kissing without either person's being aware of the infection. At least another illness, herpes, is visible during outbreaks—although the virus is always present, lingering in nerve cells under the skin. *Herpes simplex* 1, often shortened to HSV-1, is one of the ninety-plus members of the vile *Herpes* tribe—which at least (misery loves company) plague many other creatures besides human beings. Several triggers can lead to an outbreak of herpes—physical stresses such as loss of sleep, emotional trauma, constipation, the presence of other infections such as a cold, and even prolonged exposure to ultraviolet radiation in sunlight. Shakespeare is referring to this formidable barrier to kissing when Mercutio holds forth about dreams to Romeo before the masked ball. The Mab to whom Mercutio refers is a legendary Celtic imp who leads her fairyfolk dancing across our bodies while we sleep:

> *O'er ladies' lips, who straight on kisses dream*
> *Which oft the angry Mab with blisters plagues,*
> *Because their breaths with sweetmeats tainted are.*

Despite Mercutio's cautionary mention of the risks of kissing, at the ball Romeo and Juliet, having just met, engage in bawdy wordplay about kissing and then share their first kiss. Warnings have no effect on the rest of us, either. We keep kissing, just as we keep engaging in sexual intercourse. As Freud points out, "a man who will kiss a pretty girl's lips passionately may perhaps be disgusted at the idea of using her toothbrush. . . ." Incidentally, HSV-2 is a closely related virus that attacks the genitals and is communicated through sexual intercourse. The two viral cousins behave similarly and can even invade each other's turf when the conditions are right. Oral sex can convey either. Speaking of the two kinds of herpes that infect humans, the science writer Wayne Biddle remarked, "The fact that one is a social nonissue and the other a scandal is a perfect example of how our reaction to disease reflects cultural values." So far—to use the optimistic qualifier of our medicinal era—it is impossible to cure herpes once the virus is inside the body, although outbreaks can be reduced in their duration or intensity with medication.

Nowadays, while marketing sexual appeal rather than health, countless mouthwashes, toothpastes, and breath fresheners preach that we are likelier to be kissed after we buy their products. Obviously "French" kissing can convey germs such as those that cause mononucleosis, and recent studies prove that it can also transmit periodontal pathogens. And not even this sort of tongue-play is unique to us; our kissing cousins don't keep their osculation chaste, either. Bonobos kiss openmouthed with tongue action worthy of Hollywood. The activity often called in English French kissing was once called Italian kissing in France. It is also sometimes called soul kissing. As the vehicle through which we breathe, and supposedly the nozzle through which various gods spiritually inflate their little clay dolls, the mouth has always been seen as a meeting point of the soul and the body. The American actress Theda Bara embodied the fear associated with a kiss in the 1915 silent *A Fool There Was:* In a long kissing scene, the vamp(ire) begins the process of stealing a man's soul.

For a while, beginning in the early 1960s, the new technique of mouth-to-mouth resuscitation was given a boost in the popular imagination with the term *kiss of life*.

The arts reflect our osculatory obsession. Where would love songs be without kisses? Ella Fitzgerald sums up our individual discovery of the pleasures of kissing when she asks the title of George and Ira Gershwin's song "How Long Has This Been Going On?" Sarah Vaughan admitting that she feels smoochie, Billie Holiday relishing the prelude to the big moment—and that's just jazz. Every other genre has its kiss-drenched songs. Rock even has a *band* named KISS.

Kisses have been recorded on celluloid ever since an Edison tidbit in 1896—and even that one, inevitably, left the censorious hot and bothered. In Garbo's last silent, *The Kiss*, an adulterous meeting of lips seals the main character's fate. Several films have promoted actual or symbolic kisses to title roles, from the Cole Porter–scored *Kiss Me Kate* through Robert Aldrich's nuclear-noir *Kiss Me Deadly* and the very 1980s *Kiss of the Spider Woman*. Hot Lips Houlihan in Robert Altman's *M*A*S*H* earned her nickname. There have been some memorable screen kisses that helped push the limits of what society accepted among lovers. In the late 1960s, in an episode of *Star Trek*, white Captain Kirk and black Lieutenant Uhura chastely kissed, and some TV stations in the South refused to air the show. In 1971, a year of osculatory milestones, twenty-year-old Bud Cort kissed octogenarian Ruth Gordon in *Harold and Maude*, and in John Schlesinger's *Sunday, Bloody Sunday* two men kissed on-screen.

Nor do photographers ignore kissing. How many reprints have there been of Eisenstaedt's photograph of the unidentified sailor kissing a woman in Times Square on VJ Day? Painters and sculptors' immortalized kisses are everywhere you turn—Klimt's sequined contortionists, Chagall's airborne lovers, Brancusi's interfacing stone couple. Rodin's *Kiss* convincingly heats up a chunk of bronze. Some voodoo stores sell candles reminiscent of

Rodin's famous lovers; the kissing wax figures' simultaneous melt-down promises to inspire a similar ambition in the target of your lust.

Poets and novelists have bragged about the kiss as if they invented it. Petronius, citing Epicurus's contention that love gives life meaning, says flatly, "Wisdom is a kiss." At times kissing is so pleasurable that we simply do not want to stop. In one of his many obsessive appeals to the woman he disguised under the name of Lesbia, the Roman poet Catullus begs his lover for a hundred kisses, then a thousand, then more and more until they lose count in a "kissing-maze." Seventeen centuries later, Robert Herrick gradually increases the number of kisses he seeks from his beloved, until he reaches three million and suggests starting over. And three centuries further along, Louis Armstrong echoes the sentiment. In the fairy tale of Briar Rose, a kiss has the magical power to awaken her from a supernatural sleep. Some versions of the Pygmalion myth have the statue coming to life when the smitten ruler kisses it. In *Coriolanus* the returned king asks his wife for a kiss as long as his exile and as sweet as his revenge, and assures her that his lips "hath virgin'd it e'er since" the last time they kissed. Marlowe's Dr. Faustus begs Helen to make him immortal with a kiss, and even the prudish Robert Frost describes "love at the lips" as a touch as sweet as he could bear. "You must remember this," sings Dooley Wilson in *Casablanca*. Oh, we do.

Sometimes, because of its popularity and the resulting clichés that float on the surface of our thinking, kissing short-circuits the critical faculties. E. E. Cummings perpetrates a dreadful metaphor when he describes a woman's "profound and fragile lips" as the vehicle through which the feet of April come into the, God help us, meadow of his soul. Such excesses inflame cynics. Curmudgeon-at-large Ambrose Bierce, in his *Devil's Dictionary*, growls that *kiss* is a word that poets invented to rhyme with *bliss*.

In *The Guermantes Way*, Marcel (if we are to give to the narrator the same name as the author of the book) is preparing to kiss the young Albertine, about whom he has fantasized for hundreds of pages. Proust approaches kissing with his usual obsessive concentration, the microscopic analysis that José Ortega y Gasset describes as "deliciously myopic." Finally Marcel is admiring Albertine's pink cheeks and the way that they flow into "the first foothills of her beautiful black hair." Wading through Marcel's thoughts in this long section, a reader would be forgiven for impatiently exclaiming aloud, "Oh, for God's sake, *kiss* the woman!" Yet Proust does provide some interesting thoughts about kissing. He begins by pointing out that "man" doesn't really have a tool designed for this act.

> For this absent organ he substitutes his lips, and thereby arrives perhaps at a slightly more satisfying result than if he were reduced to caressing the beloved with a horny tusk. But a pair of lips, designed to convey to the palate the taste of whatever whets their appetite, must be content, without understanding their mistake or admitting their disappointment, with roaming over the surface and with coming to a halt at the barrier of the impenetrable but irresistible cheek.

Considering that they are pinch-hitting for a missing ideal organ, lips serve rather well for kissing. Our lips' concentrations of sensitive nerve endings almost automatically seek out a match in another person. We can safely assume that many of us exist today because a kiss between a man and a woman heated up and led to other activities. At this very moment, all over this wet ball of rock spinning through space, human beings are kissing—on subways and on horseback, in elevators and in grass huts, on airplanes and on yachts, on beaches and in yurts, and of course in the backseat of countless automobiles. At this moment a middle-aged couple in Soweto are granting each other a chaste peck before turning out the light, and beside the Yangtze an elderly woman is bestowing a farewell kiss upon her dying husband, and in Oregon teenagers are leaning closer and closer until their lips touch for the first time.

THE NECESSARY AND AMBIVALENT TONGUE

Many have fallen by the edge of the sword:
but not so many as have fallen by the tongue.

ECCLESIASTICUS

Although the tongue is inside the head, guarded by the decadent lips, it is highly visible and important in our lives. Like the hand, the tongue caresses the world, and likewise its virtues blossom in our fertile brains into symbolic images. A peninsula, a bell's clapper, a leaping flame, a shoe's flap of leather—all are called a tongue. Holy Rollers speak in tongues. A shy man loses his tongue, and a critic has a sharp one. Consciousness embodied: With our animal tongue we express the uniquely human images glowing in our ancient brains.

Because this versatile muscle shapes verbal communication, throughout history and around the world—especially before written language became widespread—despots and criminals have cut out the tongues of victims and witnesses. An example of this terrible punishment appears in Greek mythology. King Tereus tricks Philomela, his wife Procne's naïve sister, and rapes her. As Ovid tells the story, Philomela cries out to her father, to Procne, and to the gods above, but no one hears her. She swears to Tereus that her voice "shall fill the woods / And move the rocks to pity." The king binds Philomela and approaches her with his sword. Wanting to die, she bares her throat, but he grabs her tongue with tongs and cuts it out with his sword. Eventually Philomela and her sister mete out vengeance on the rapist. As the sisters flee, the negligent gods finally intervene and turn them into birds to hasten their escape. (In the original Greek story, Philomela becomes a swallow and Procne a nightingale, but later Roman poets reversed the names; by the early nineteenth century, Keats naturally referred to the nightingale as Philomel.)

Sometimes the tongue ceases to form words not because it has been removed but because the words are too terrible to speak aloud. In her dreamlike, heartbreaking narrative *Under the Tongue*, the Zimbabwean novelist Yvonne Vera writes about Zhizha, who is raped by her father, Muroyiwa. After the crime, and after her mother is jailed for killing her father, Zhizha stops speaking. Throughout the novel Vera employs a consistent metonymy: The tongue represents speech itself. "I know a stone is buried in my mouth, carried under my tongue. My voice has forgotten me. . . . A word does not rot unless it is carried in the mouth for too long, under the tongue." It is the wordless cry of silenced women all over the world.

Although the tongue alone cannot produce speech, this kind of personification runs through culture and conveys our dependence upon the organ. "O necessary Tongue!" exclaims Thomas Adams in his 1619 work *The Taming of the Tongue*. "How many hearts would burst, if thou had not given them vent!" As Adams's own title indicates, although we know that the tongue communicates the yearning of the heart, we also fear its candor and its power. Consider the titles of some other books published in the same century—*The Poysonous Tongue, A Bridle for the Tongue, The Government of the Tongue*. As Thomas Adams says, "The eye, the eare, the foote, the hand, though wilde and *unruly* enough, have been *tamed, but the tongue can no man tame.*" At about the same time, in his *Collection of Emblemes*, the artist George Wither portrayed the tongue's willful independence by drawing it as a grotesque winged creature independently rising upward from the earth. The caption reads "*No Heart can thinke, to what strange ends, / The Tongues unruely Motion tends.*" But surely the fiercest denunciation of this maligned organ appears in the Bible, where the apostle James shrieks in his hortatory epistle to the twelve tribes of the Dispersion, "And the tongue *is* a fire, a world of iniquity: so is the tongue among our members, that it defileth the whole body, and setteth on fire the course of nature; and it is set on fire of hell."

Enough of the tongue's iniquities. It does more than speak.

◎

In the early sixteenth century, Erasmus disparaged the tongue as merely a "flabby little organ," but the great humanist also exclaimed over its virtues and vices, "O ambivalent Organ." The tongue is a talented and powerful part of the body, and no wonder we are ambivalent about it. Along with the lips and larynx and lungs, it forms words and other sounds. Yes, we might reply to the apostle James, the tongue promises and reassures and lies; it gossips, pronounces love and lust, orders executions, shapes compliments and reprimands. Yet in its innocent animal way it also performs its original tasks of tasting food and aiding in chewing and swallowing, bringing us nourishment and carnal delight. All creatures experience the world through their bodies and use their bodies to respond.

There is as wide a variety of this organ in nature as of any other body part. The tongue is seldom noteworthy in birds, but the flamingo is cursed with one so muscularly tasty that Roman emperors served them by the bowlful. Hummingbirds hover militarily while unfurling their tongues to French-kiss flowers. In the nineteenth century, countless American buffalo were massacred, their corpses left to rot on the plains while their tongues met those of human beings. As if being practically invisible were not enough for a lizard, with muscles unique among vertebrates a chameleon can also shoot its tongue more than twice the length of its own body. Overheated dogs appear to be laughing because they release heat through their tongues. Thanks to widespread bias against snakes, their forked tongue represents dishonesty when actually it is merely reading the air's honest molecules. Yet basically the same tool that a deer uses to lick salt a toddler employs to sculpt a cone of ice cream. Every baby, every puppy, hurries to lick the world. Lingua franca: a common tongue.

Although our tongue lacks some of the talents of our fellow creatures', it makes up for it by framing one of life's great inventions and delights, the faculty that worried the authors of *The*

Poysonous Tongue and *A Bridle for the Tongue*—language. The very word refers to the tongue itself, which also hides in *lingo,* *glossary, linguini, bilingual, cunnilingus,* and *glossolalia* (speaking in tongues). Our word *tongue* traces back to Middle English *tunge* and *tonge.* Many cognates—the Old High German *zunga,* for example—are similar. All derive from the Latin *lingua,* and *language* from its descendant, the Old French *langue.* About the ancient habit of linking this organ with the meaningful sounds it helps produce, philologist Eric Partridge remarked, "The 'tongue-language' dualism is as widespread as it is natural." Carla Mazzio, an American literary scholar, goes further: "Because 'tongue' . . . also means 'language,' the very invocation of the word encodes a relation between word and flesh, tenor and vehicle, matter and meaning."

The tongue's diverse accomplishments begin with the cooperation of intrinsic muscles—vertical, longitudinal, transverse—and elaborate anchoring to extrinsic muscles. Its receptors for various tastes are segregated. Taste buds on the sides detect sourness; saltiness registers along the edge. With the tip of the tongue we know sweetness immediately, but bitterness must reach the back of the mouth before we recognize it. The rounded elevations on the tongue are called *papillae,* as are similar nipplelike protrusions on the ridges of the fingerprints. Just as the papillae on the fingertips perform two jobs—grasping and providing founts for the sweat glands—so the papillae on the tongue both move food and house the taste buds. The buds line the sides of the papillae. Some show up on the walls of the pharynx (the throat and the back of the nose) and the roof of the mouth, but most taste buds are on the surface of the tongue, especially toward the back. The buds have an opening called the taste pore that provides access to its internal taste cells. Through such microscopic chemistry we savor chocolate and peaches.

And the tongue is master of more than just taste. It touches in its own particular way, too. The clever tip ferrets out bits of food from crannies in teeth and fetchingly swipes whipped cream off lips. It meets another tongue, cautiously or with abandon, and causes all sorts of things to happen elsewhere on the bodies involved. The tongue can detect on its wet surface a single poppy seed or grain of salt. A few people can twist this sensitive muscle to knot the stem of a cocktail cherry, and usually the same individuals can roll it into a tubelike shape. Those with less skilled tongues can comfort themselves that such a talent seems evolutionarily pointless.

DON'T READ MY LIPS

The art and science of lipreading demonstrates the way that the lips and tongue shape the sounds we use in communication. Many deaf people learn to read lips. This ability can be helpful to an espionage agent, although it is less relied upon now that we possess ever more sophisticated amplification devices. Lipreading made for some good scenes in old movies, with the agent watching through binoculars: "What's she saying?" "She seems to be singing along with the radio." For all the difficulty of watching a foreign film while reading subtitles, the brain seems to find this task less confusing than the misalignment of lip movement and sound in dubbed versions. Long ago we evolved the habit of looking at the speaker's mouth movements when listening.

Because mobility is part of the very definition of lips, its absence can be noteworthy in itself. With practice many people can produce convincing speech without noticeably moving their lips. The term *ventriloquism* records the venerable history of this practice. It means "belly-talking." In ancient Greece and Rome, ventriloquism was one of the secret talents in the arsenal of diviners. A skilled medium could hoodwink the gullible into believing that spirits were communicating through his divinely favored belly. Shamans around the world have been using this con-man method for thousands of years. Occasionally the practice backfires. Consider the case of Elizabeth Barton, the notorious "Holy Maid of Kent" in sixteenth-century England. Employing her ventriloquial talent, Barton claimed that through her abdomen dyspeptic spirits were prophesying against the impending second marriage of Henry VIII. The king's we-are-not-amused response was to hang Barton and her chatty torso.

Demoted from spiritual telephone to entertaining trick, ventriloquism thrives even today. Quite a few people make a living primarily through their ability to speak without moving their lips. Now as always, this is accomplished with methods that require considerable practice. Avoiding, as much as possible, difficult lip-involving consonants such as *b* and *p* and *m*, ventriloquists must train themselves to approximate these sounds with clever movements of the hidden tongue. Professionals maintain that, although technique can be improved, the adroit tongue is innate. Of course, like conjurors, ventriloquists also employ misdirection. They distract viewers from their mouth movements by giving the dummies realistic gestures and by turning to face the dummy when listening to it "speak."

It is part of the charm of the human race at its goofiest that we are amused by such tricks. In the 1940s the great ventriloquist Edgar Bergen, creator of smart-aleck dummies Charlie McCarthy and Mortimer Snerd, was popular on radio. His success in the medium is a tribute to his genius for creating believable characters and amusing dialogue, because radio removed the very point of ventriloquists: the opportunity to watch their lips *not* move when they perform. On radio, who knew whether Bergen's lips were moving? The popularity of ventriloquism is also a reminder that, carried along by the momentum of the human imagination, the cultural history of the body can move far from its starting point with a natural phenomenon—in this case with those mobile, versatile, sensitive lips.

PART TWO

The Weight
of the World

CHAPTER 7

Arms and the Man

That muscles have become status symbols signifies that most jobs no longer call upon bodily strength: like tans, they are an aesthetic of the obsolete.

REBECCA SOLNIT

Greek mythology features many stories inspired by the powerful muscles in our arms and shoulders. Take, for example, the grand old myth of Atlas. The son of Iapetus and brother of Prometheus, he is a gigantic Titan who stomps around on Earth in the days before the Olympian deities are fully in control. Because Atlas fights on the losing side in the battle between the giants and the gods, Zeus sentences him to carry the vault of the heavens on his massive shoulders. Part of the fun of mythology is the way that deities—Greek, Christian, Navajo, Khmer—personify, literally embody, all our physical attributes, exaggerated to the comic-book level of superheroes and supervillains. Naturally, the king of the gods would decide that if something needed to be held up, the best tool to do the holding would be the animal strength of the shoulders and arms—the "back" in the backbreaking toil behind all of preindustrial civilization. What else could support the weight of the world?

The giant's story inspired a diverse legacy. The Atlas Mountains in northwestern Africa are named for this fundamental armature, whose home was considered beyond the horizon of the Greek world. More to the point of our story of the body, and to the ways that its cultural history reflects its natural history, are two fossil terms. In architecture an atlas is a sculpture of a human figure used as a supporting column, otherwise known by another Greek name, *caryatid;* and anatomists call the first cervical vertebra, which supports the heavy and celestially minded head, the *atlas vertebra.* A bound collection of maps is called an atlas because early volumes often included a picture of the weary giant holding up the globe itself rather than the sky. Incidentally, in case you have several of either around, the plural of *atlas* is *atlases* but more than one supporting column are *atlantes.*

The story of Atlas is reminiscent of the tales of Hercules, son of Zeus and Alcmene, and in some myths the two strongmen cross paths. Although he was also cunning and resourceful, in his famous twelve labors Hercules relied upon his strength in many ways. He had to kill the Nemean Lion and the Lernaean Hydra, carry the Erymanthian boar on his sturdy shoulders, and even venture into the underworld to capture the dog Cerberus. To this day we describe arduous tasks as herculean. There are many other myths about the strength of this part of the body. Ares, the god of war, is irresistibly strong. So is Amphissus, the son of Apollo. Just before the fall of Troy, Aeneas carries his father Anchises to safety on his own shoulders.

None of these figures, however, had as challenging a burden as the giant who supported the heavens. His reputation was familiar enough to inspire, as late as the twentieth century, a curious incarnation of Atlas as tower of strength. In 1894, thousands of years after the first drunken Greek shepherds gazed at the reeling stars and wondered who held them up, an infant was born in Italy. Angelo Siciliano was brought to the United States as a child and grew up in poverty in New York City. Later he described a key event in his life, during a time when he was the proverbial ninety-seven-pound weakling.

One day I went to Coney Island and I had a very
pretty girl with me. We were sitting in the sand. A big,
husky, lifeguard . . . kicked sand in my face. *I couldn't
do anything* and the girl felt funny. I told her that
someday, if I met this guy, I would lick him.

The italics are Siciliano's own, although by the time he
wrote these words he was using a different name. In an unlikely
anecdote, he explained how he chose it. Soon after the lifeguard
kicked sand in his face, Siciliano had no girlfriend to wrap his
scrawny arm around. Supposedly, as he walked home dateless and
embarrassed, he encountered a poster of Atlas supporting the
world on his shoulders. Later he claimed that at this very moment he resolved to rebuild his body into a force to be reckoned
with. He was so successful in turning himself into a strongman
that he wound up working as a sculptor's model, and he decided
that he could help other underdeveloped (and therefore less-
than-masculine) men. He renamed himself Charles Atlas and
progressed from Coney Island sideshows to vaudeville. In time he
developed a system of muscle-building that he called "dynamic
tension." By 1922 *Physical Culture* magazine was calling him the
"World's Most Perfectly Developed Man."

"Are You a Redblooded Man?" taunted the advertisements
for the Charles Atlas system. The ads compared strong men to
tigers and stated flatly that it was these brawny Tiger Men who
"grab everything they want" and "win the battles" and, as a consequence of the modern way of keeping score in the Darwinian
struggle for existence, have "bulging bank accounts." Frequently
such promotions appeared in magazines and comic books that
were aimed at adolescent boys, most of whom were already being
indoctrinated in sports and other social cues of red-blooded manhood. The boys' struggle for self-definition inspired them to save
their pennies and send away for Mr. Atlas's patented method of
masculinity enhancement. Whether their newfound strength led
to their own bank accounts' bulging is not recorded, but it is a

safe bet that Charles Atlas seldom bounced a check. He managed to build a flourishing business during the Depression, and by the Second World War he was the proud owner of the most successful mail-order company in U.S. history.

As Charles Atlas demonstrates, arm strength plays an amusing and predictable role in the cultural history of masculinity. In his wimpy early days, the young Angelo Siciliano spurred himself on toward greater physical development with an inspirational photo of an immigrant Prussian bodybuilder named Eugene Sandow. At the turn of the century, Sandow was already acclaimed as "the strongest man in the world," the most beautiful example of the (white) male body. Although to admit so in public would have undermined his pretense of originality, surely Sandow himself was influenced by the exercise movement founded by Friedrich Ludwig Jahn to restore a sense of empowering masculine strength to German men defeated in the Napoleonic Wars. The era, like most in history, was rife with posturing men. Through carefully staged demonstrations, usually recorded by photographers, Theodore Roosevelt was always advertising his vigor to inspire the public with his manliness. Houdini emphasized not only his agility and expertise but also his sheer strength. To quote cultural historian John F. Kasson, "Supposedly a biological category, manhood was also a *performance*." This observation is still true, of course. During the 1984 U.S. presidential campaign, Ronald Reagan demonstrated his usual commitment to substantive debate when he said of Walter Mondale, "I'll challenge him to an arm wrestle anytime."

Presidential bravado reminds us that often power of any kind is compared to the human arms and shoulders. Nor is this symbolism always masculine. Several of the strong heroes are female. Two of Athena's many incarnations were Athena Alalcomeneis, "Powerful Defender," and Athena Alcis, "the Strong." When we say that an activity is strenuous, we are remembering the Roman goddess Strenua, who in difficult times lends some of her own great strength to weaker mortals. During World War II, women and the populace at large were inspired by the figure of Rosie the Riveter, whose strong arms promised that she could handle her assigned tasks as well as any man could. Reproductions of one poster, featuring Rosie with her arm upraised and her hand forming a fist, bore the caption WE CAN DO IT! It still flourishes in Women's Studies departments.

As an active agent and the supporter of the versatile hands, the arms and shoulders are also involved with other kinds of imagery. Well before contemporary Americans were talking about the long arm of the law, Herodotus was saying that the king has a very long arm. Some scholars claim that Plato, whose real name was Aristocles, derived his better-known name (the Greek *plat* meant "broad" or "flat") from his wide forehead or wide learning, but others think it was because of his broad shoulders. The symbolism of arms can be more complicated than one might imagine. We raise an arm in class merely as a flag to attract attention—the way that we hail a taxi—but we also raise an arm and hand when we testify in court. Enceladus, the son of Tartarus and Gaea in Greek mythology, had a hundred arms, as did Gyges. The multiple arms of Hindu deities embody their many tasks and frequently carry specific symbols. Two of the arms of Siva Nataraja, the Lord of the Dance, support the flame that represents destruction balanced against the drum that represents creative rhythm.

The opposition in these two symbols also acknowledges that a primary atrribute of arms is the way that they collaborate as a unit. In 1923 Mustafa Kemal, the founder and first president of the Turkish Republic (and therefore called Atatürk, "the Father of the Turks"), used the paired cooperation of the arms and legs in a memorable statement on behalf of the status of women in a more balanced populace: "If a society contents itself with modernizing only one of the sexes, it will be weakened by more than half. . . . If in a society one limb is active while another is inert, that society is paralyzed."

Fortunately, our arms usually work together. Human beings are tetrapods. Although taxonomists continually refine these classifications, in general tetrapods include mammals, birds, reptiles, and amphibians; that is, all vertebrates who possess two paired appendages—except for fishes, whose evolutionary line traces a different ancestry. (Such creatures as snakes have since lost their limbs, but their ancestors were tetrapods.) This versatile arrangement of limbs evolved during roughly 400 million years, a family tree that is traceable all the way back to the Crossopterygian fishes of the watery Devonian period. There is only one known living form of these creatures, the famous "living fossil" coelacanth, considered extinct until discovered in 1938 calmly swimming along in its outdated vehicle of a body like an old man driving his grandfather's car. Technically, like sharks, they are not fish.

We have several other bodily features that reflect this aspect of our evolutionary legacy. Like these marine forefathers, we possess a central skeletal axis, a bony skeleton, and paired limbs. Any science museum's public collection of skeletons reveals the similarity of structure between ourselves and our relatives, including those who seem at first to share little family resemblance. When you compare the bones under the skin, you find that the customized garden tools forming the paw of the star-nosed mole are remarkably similar to the bony armature supporting the wings of the blue-footed booby and the greater horseshoe bat. Hidden in the fins of the blue whale, which has moved back into the water after its tetrapod ancestors adapted to land, are finger bones not too dissimilar from those in the hand of the stumptailed macaque. The whale lacks a hind limb, but the forgotten bones are still there, buried under skin and blubber. Not only are our limbs analogous to those of other creatures; the fore- and hind limbs are analogous to each other. On ourselves the upper arm corresponds to the thigh, the elbow to the knee, the wrist to the ankle, the palm to the sole, and the fingers to the toes.

In our case, as in that of other creatures from squirrels to chimpanzees, the limbs have diverged for different tasks. When you pick up your luggage, you are using an ancient lifting device. The evolution of an upright stance freed the hands and arms from the demands of quadrupedal transportation and allowed them to evolve for manipulation. However it came about, this advantage is apparent every time you carry groceries home from the market—and celebrated in every chin-up. Brachiation, the arm-swinging form of locomotion that propels our primate cousins through tropical canopies, is the ancestor of our own arm and shoulder movements, which we demonstrate every time we climb trees or play volleyball. The brachial plexus is the network of spinal nerves, running through cervical and thoracic vertebrae V to IX, that innervates our forelimbs. These control-wires from the spine direct the trapezius, deltoid, pectoralis major and minor, and other muscles, providing the strength that inspires myths of Atlas and Hercules.

As bioengineer Steven Vogel says in *Prime Mover*, his natural history of muscle, "Both the pyramids of Egypt and the Great Wall of China represent muscular accomplishments as much as they do social and technological ones." Muscle takes up almost half our average body weight and permits and directs every movement. The hidden bones are deaf to the brain's instructions; they are tugged like the segments of a marionette by the muscles, which in turn are commanded by the most mysterious and powerful member of the body politic, the nerves. The dancing skeletons of painting and fable are horrific partially because they lack the covering that makes bones move—flesh, meat, muscle. "Our muscle," writes Vogel, "differs only a little from that found elsewhere in the animal kingdom. . . . It powers ant and elephant alike, so alike that only a trained eye can see the subtle differences between their muscles when bits are viewed under a microscope." The similarity between the gastrocnemius muscles in our legs and frogs' legs—we need to walk and climb, and they need to jump—leads Vogel to point out that "frogs have fine calves." It is no coincidence that these calves are edible. And it is not coincidental that chickens have tasty wing muscles and cows have delicious thighs. Most cultures that eat meat savor juicy and nutrient-rich muscle. Human beings even eat the muscles of their own kind. Anthropologists'

studies of primitive societies indicate that nutrition-based canni-
balism has usually focused on muscles, the best flesh.

We owe a great deal to the anonymous people who coined the
vivid ancient words that survive today in later mutations. A new
term does not coalesce from the zeitgeist by itself; a first inventive person must speak it aloud to a listener. Someone with a
good imagination coined the word *muscle*. Watch your biceps as
you raise and lower your arm, and you will observe that the muscle tightens into a lump and then stretches out again. Long ago
this movement reminded somebody somewhere of either a
mouse's scurry or its ability to flatten itself to crawl into tight
spaces. In time the ancient Latin word for a mouse, *mus*, became
our word *muscle*. Commenting upon this etymology, the novelist
Paul West noted that "if you are unlucky enough to be bitten by
a black mamba, your leg muscles will ripple and buckle in precisely this '*musculus*' way, but surely the Romans never saw that."
    Our arms have inspired many such terms. Because of the
construction of their meaty shoulders and arms, the great apes
are better equipped for energetic striking and throwing than are
monkeys. We have used this talent to hurt our fellow human beings, creating spear launchers and arrows to surpass the strength
of our own arms. From this enhanced musclepower we derive a
powerful metaphor. "*Arma virumque cano . . .*," intones Virgil in
the opening line of his first-century-B.C.E. epic about Aeneas and
Dido. John Dryden's lively rhyming translation is still the most
quoted in English:

> *Arms, and the man I sing, who, forced by fate,*
> *And haughty Juno's unrelenting hate. . . .*

    In contrast, Robert Fitzgerald begins his 1980s translation of
the *Aeneid* with "I sing of warfare and a man at war. . . ." The difference in translation highlights the symbolism. Dryden was not
referring to literal arms any more than Virgil was; as Fitzgerald

makes clear, Virgil was referring to weaponry. His word choice, however, exemplifies a nice connection between the different usages of the term *arm*—as both a limb supporting a hand and a weapon wielded by a real arm. We are used to these two meanings. We speak of armor and armed forces and unarmed bystanders, and we know that Hemingway's novel *A Farewell to Arms* is not about amputation or the Venus de Milo. Some scholars argue that the Latin *arma*, the root of both *arms* and *army*, may well derive (via *armare*, "to equip with arms") from *armus*, the root of the limb's name. This is an appealing conjecture, because throughout history and around the world—from the myth of David's conquering Goliath to Mark David Chapman's pistol aimed at John Lennon—weaponry arms are brandished by human arms.

This talent for harming works both ways. The loss of an arm has always been a risk when enemies armed themselves against you. Is this danger related to our fascination with the illusive amputations and decapitations that magicians perform onstage? Unfortunately, the reality never ends with the lost limbs restored. The French novelist Émile Zola, ever committed to *le réalisme*, wrote with admirable bluntness about the panoply of ills that torture humanity. He never blinks in his documentation—the anguish of the faithful sick in *Lourdes*, the broken victims of the Franco-Prussian War in *The Debacle*. The latter novel, which was published in 1892, portrays the horrors of war without sentimentality or posturing, by concentrating on the suffering of the soldiers and civilians rather than on the machinations of their leaders. Zola includes scenes of triage and emergency surgery that graphically depict amputations either without anesthetic or assisted only by unreliable chloroform. Zola's description of discarded limbs powerfully conveys the casual waste of life: "And at the feet of the dead a jumble of legs and arms had piled up, everything that had been clipped away, everything that had been lopped off on the operating tables, the sweepings of a butcher's shop. . . ." A soldier jolts awake to see his arm being carried away and looks at his own bloody limbless shoulder. He begins to weep and cries out, "What am I bloody well going to do with myself now?"

It is the cry of victims throughout our war-torn history. Nowadays one answer (although hardly sufficient recompense) is artificial limbs. New bionic arms testify to the complexity of our simian nerve systems. One prosthesis, developed by David Gow and others at the Edinburgh Royal Infirmary's Bioengineering Centre, attaches to a stumpless shoulder with Velcro straps and uses electrodes to detect the electricity from the nerves and move the arm in response. One recipient, a Scot named Campbell Aird, said of his bionic replacement, "To make it move, I just flex my shoulder and back muscles." These devices are still in their infancy, and not everyone can wear one. With or without bionic help, quickly developing the aplomb that so impresses those of us who have escaped this misfortune, people adapt even to the loss of a limb.

We are used to the balanced symmetry of two arms, and also to their relatively standard proportions. Dwarfs look odd to other people not just because they are shorter but because they seem disproportionate. Certainly there are mean ratios that we assume, noting any variation therefrom. E. H. Gombrich tells a story along these lines in his classic *Art and Illusion*. A woman examining a painting in Matisse's studio complained to the artist, "But surely the arm of this woman is much too long." And Matisse replied, "Madame, you are mistaken. This is not a woman, this is a picture." The woman saw the artistic representation of the arm in terms of its reality, just as many people see aspects of the real arm—such as strength—in terms of its myths. Perhaps both the woman and Matisse gestured toward the painting. If so, they waved not a symbol, an ideal, or an abstraction but the combination of muscles, bones, ligaments, tendons, and nerves that make up that splendidly real tool, our tetrapod foreleg, our anthropoid shoulder, our human arm.

CHAPTER 8

The Monkey's Paw

Man could not have attained his present dominant position in the world without the use of his hands, which are so admirably adapted to act in obedience to his will.

CHARLES DARWIN

A FREE HAND

The distinguishing mark of man is the hand, *the instrument with which he does all his mischief.*

SNOWBALL,
in *Animal Farm,* by George Orwell

The English artist Edwin Landseer, remembered mostly for his sentimental paintings of animals, was also famous for his ambidexterity. Sometimes witnesses observed him drawing one animal with his left hand while drawing another with his right. In 1840 the dexterity of Landseer's own hands paid tribute to those of a fellow artist, violinist Niccolò Paganini. Landseer captured the musician's passionate energy with what appear to have been lightning strokes—perfectly crosshatched sideburns, the wrinkles of sleeves, and most of all the barely drawn tender authority of the hands. One, a jumble of fingers, holds the bow upright; the other grips the neck and caresses the strings with a firm, light touch, forefinger upraised, little finger delicately stretched to complete a chord. In this work both artist and subject are monuments to the astonishing capabilities of the human hand.

We are not all virtuosos, but our multitalented hands shape our lives. The first time a baby wraps her tiny hand around her mother's finger, she begins a lifetime of holding on and letting go. Although most of us tend to take our own hands for granted and marvel at those of violinists or magicians, we rely constantly upon their dexterity and sensitivity. Think how casually we tie shoelaces and sign our names. We caress loved ones, pick up contact lenses, throw Frisbees. We can perform heart surgery or fire a handgun. In one of his *Far Side* cartoons, Gary Larson expressed the role of the hand in our dominance over other animals by portraying a laboratory full of dogs agonizingly trying to decipher the "Doorknob Principle."

A few minutes of watching your own hands will remind you of the versatility that usually we take for granted. Imagine that you are walking from your car to the front door of your apartment building. It is raining, and with your right hand you hold aloft an umbrella, in the squeeze grip that we also use for such actions as grasping a hammer. In your left hand you carry a briefcase, with the simple hook grip that our cousins employ to brachiate, or swing from branch to branch. Along the way you pass your neighbor, who is holding a cigarette between her first and second fingers in a scissors grip. You put down your umbrella and reach for your apartment key. At this point you remember recent studies indicating that *Homo sapiens neanderthalensis* may have been better at power grips such as the one we use to throw a rock, but that our own ancestor, early modern *Homo sapiens sapiens*, was better at precision grips. At the door you employ two variations of a familiar precision grip that specialists call a two-jaw chuck. To take the house key out of a pocket, you grasp it between thumb and index finger, but to insert the key into the lock and turn it, you hold it between the thumb and the *side* of the index finger. In the kitchen you twist the lid off a jar with a disc

grip, which is related to the spherical grip you will use this week-
end to throw a softball. Finally you use yet another type of hold,
called a five-jaw chuck, to set the lid aside.

To perform such a variety of functions, hands, like a computer application, require a lot of space on the hard drive. During
a career spanning more than half of the twentieth century,
Wilder Penfield devised a map of the body that reveals how
much the brain values the hands. Penfield was an American-
born neurosurgeon who emigrated to Canada and founded the
influential Montreal Institute of Neurology. His diagrams portray
the body not as it looks to us but with each area in proportion to
the amount of the brain's cerebral cortex devoted to it. Penfield
mapped the motor cortex, which monitors movement, and the
somatosensory cortex, which monitors sensations of touch, and
called these diagrams *motor homunculi* and *sensory homunculi.*
These representations have now been mapped on paper, in three-
dimensional structures, and on computers. In all models two
areas dominate, cartoonishly distorting the body. The head in
such maps is hugely disproportionate, especially the face and
most especially the eyes and mouth, and the hands are gargan-
tuan. The genitals receive more attention than their size relative
to the rest of the body might predict, but they do not compare
with the head and the hands. Like government funding, the
brain's allotment of resources demonstrates priorities.

The brain's neurological budget is administered by the fingers' local offices. Wonderfully tiny nerve endings are, as the English polymath Jonathan Miller says in *The Body in Question,*
"embroidered like millions of seed-pearls throughout the fabric
of our body." However, as a Penfield diagram proves, these nerves
cluster in particular areas. Many neurophysiologists speculate
that *Homo sapiens* has lost the richly innervated skin that characterizes other animals; we have sensitivities truly comparable to
theirs only in our erogenous zones and on the tips of the fingers.
In these areas the population of nerve endings can reach thirteen
hundred per square inch—explaining Miller's remark that a Penfield map is electoral rather than geographical. On this theme of

fingertip sensitivity, the meager output of the twentieth-century Russian poet Ksenya Nekrasova includes a poem about a blind man, whom she describes as having "ten eyes in his fingertips." She wrote of his way of touching the world, "here is a man perceiving the nature of things." An American neurosurgeon and writer, Frank T. Vertosick Jr., expresses the same importance of touch in medical terms when explaining carpal tunnel syndrome (which we will discuss later): "The median nerves are lifelines to our hands, and our hands a link to the physical world. Numb hands are like blind eyes."

Another connection between brain and hands shows up in the way that babies who are exposed to sign language quickly begin to gesture in the kind of goo-goo babble that characterizes first attempts at speech. In a 2001 paper in *Nature*, Laura Ann Petitto, a psychologist at Dartmouth, published findings on this fascinating topic. "You see bite-sized nuggets of sign on the hands," she says—tidbits of language indicating that centers in the brain are primed for the acquisition of language without narrowly defining language as speech. The infants gesticulate their silent prattle in a "sign space" in front of the torso. Another Russian poet, Nikolai Glazkov, describes deaf-mutes reading poems aloud with their hands.

With so much invested in them, it is not surprising that hands are more complex than they look. They have so many nerve endings and so much brain space behind them that they seem to remember on their own. Musicians claim that after a while melodies and chord structures reside in their fingers rather than in their brains. Long after his apprenticeship days as a parliamentary reporter, Charles Dickens would dictate a letter or listen to a speech with his fingers unconsciously making little twitching gestures as if recording everything in his precise shorthand. "Our legs and arms are full of torpid memories," writes Proust in the final volume of *In Search of Lost Time*. The narrator tells of half waking and finding that his arm, with its own history nudged into recollection by a chance movement, is reaching to ring the bell for the long-dead Albertine. Raoul Ruiz beautifully visualized this disconcerting moment in his 1999 film of *Time Regained.*

One side effect of hands' survival-related sensitivity is the matter of pleasure. Consider the moment when two individuals' hands automatically seek each other. Entangled fingers can be as arousing as entangled arms and legs. One such sexy encounter occurs in Jean-Jacques Annaud's 1992 film *L'Amant* (The Lover), based upon Marguérite Duras's lyrical novel. The first time they meet, on the ferry crossing the Mekong, a teenage girl and an older but still young man sit in his car and slowly, ever so slowly, inch their hands across the seat between them—until, with all the intensity of actual lovemaking, their fingers touch for the first time. How many times has this act occurred? The Beatles were not the first to want to hold someone's hand. It is the first step toward intimacy.

Other primates also touch in this way. Jane Goodall describes a young chimpanzee, frightened by an older male, reaching instinctively for his mother's hand. Like kissing and hugging and bowing, all of which Goodall and others have observed in chimps, hand-holding is an ancient primate habit that predates human beings. Once Goodall offered fruit to a chimpanzee by holding it out on the palm of her hand. The chimp looked at her hand, then at her face, and finally took the fruit and grasped her hand. Goodall later wrote, "The soft pressure of his fingers spoke to me not through my intellect but through a more primitive emotional channel: the barrier of untold centuries which has grown up during the separate evolution of man and chimpanzee was, for those few seconds, broken down."

Because hands are so important, the handshake has evolved as a common form of greeting between friends and even strangers. As the ever articulate George W. Bush once asked, "The important question is, How many hands have I shaked?" For many politicians and businesspeople, this question is indeed important. The handshake as we know it did not become firmly established until early in the nineteenth century, and it reflects a shift in assumptions about class and rank. Its predecessor, the handclasp, dates back to ancient Rome. Varieties of handclasp included the *dextarum iunctio*, the act of a bride and groom joining right hands to pledge troth. By the 1500s there are records of the act of grasping hands' being accompanied by a shaking motion, and still it was not the handshake as we know it. Theories about the origin of the handshake must reach back into prehistory. One form of handclasp common among our fellow higher primates is the submissive offer of a limp hand to a male of higher rank. Ancient literature suggests that the act of presenting an empty hand indicated truce: I, a stranger, come to you as a potential friend. However, the handclasp of the Middle Ages was accomplished while one person genuflected or bowed to submit to the higher rank of the other. By the early 1800s, the rise of the growing middle class had a double side effect on the evolution of the handshake. Greater commerce demanded more deal-making rituals, but it also began to erode class distinctions. In our own time, handshakes are most often relatively neutral greetings.

There are many kinds of handshakes—viselike, hearty, limp, cool, sweaty, dry. An interesting example of the social significance of a *firm* handshake concerns Marcel Proust's homosexuality, which apparently he genuinely thought he was managing to hide from his peers. As Edmund White tells it in his elegant little Proust biography, the Romanian prince Antoine Bibesco described the young Proust as having "Japanese lacquer" eyes and a "dangling and soft" handshake. Bibesco attempted to teach Proust the official firm handshake that traditionally betokens red-blooded masculinity. "If I followed your example," responded Proust with his usual labyrinthine reasoning, "people would take me for an invert." White sums up this sad contradiction as "an indication of how devious the thinking of a homosexual of the period could become—a homosexual affects a limp handshake so that heterosexuals will not think he is a homosexual disguising himself as a hearty hetero—whereas he is in fact exactly what he appears to be: a homosexual with a limp handshake."

◉

Inevitably, culture reflects the brain's preoccupation. Few symbols are more common than the hand. *Handy, handle,* even *handsome* (which originally meant "easy to handle") all derive from the same familiar source. Paul extended his hand to the Galatians, and Pilate washed his of the whole affair. Many hands make light work, and bad dogs bite the hand that feeds them. You may think that you are in good hands, but if matters get out of hand you could fall into the wrong hands. You can give a hand by applauding or by assisting. "Give a man a free hand," warned Mae West, "and he'll run it all over you."

Holding West's hand in *My Little Chickadee,* W. C. Fields observes, "What symmetrical digits!" Never mind the arguments over innate versus perceived beauty; Fields knew what Thomas Aquinas knew: "The senses delight in things duly proportioned." An entire episode of *Seinfeld* revolves around the comedian's reluctance to date a woman who, while otherwise attractive, possesses oversize "man hands." Although both male and female hands vary widely in size and proportion, the hand was a unit of measure in many cultures, including among the Greeks, Egyptians, and Hebrews. During the time of Henry VIII, a decree established the official width of the hand as four inches. This is the span it still measures when denoting the height of horses.

Often the hand is employed in synecdoche, the rhetorical device by which a part stands for the whole or the specific represents the general. "All hands on deck!" summons the sailors' complete bodies, just as a suitor asking a daughter's hand in marriage probably wants the entire daughter. (Related wordplay occurs in a Dudley Do-Right cartoon in which Dishonest John abducts Dudley's true love and cackles at him, "You'll never get her back—or any other part of her!") Frequently an active agent is called a hand. The German poet Heinrich Heine grotesquely described Robespierre as "the bloody hand that drew from the womb of time the body whose soul Rousseau had created."

Photographers have been unable to resist the hand as a subject. Alfred Stieglitz repeatedly photographed the beautiful, talented hands of his protégée and lover, Georgia O'Keeffe. More recently, the American photographer Holly Wright portrayed the essential strangeness of the hands that we no longer see because we see them constantly. After photographing blurry actions such as applause, Wright began in 1985 to document her own hands. Dramatically lit in relief, close-ups of wrinkled mountainous landscapes depict our familiar appendages as new and alien.

Photographers and filmmakers are only the latest artists to memorialize the power of hands. In van Gogh's well-known drawing of a carpenter at work, the man's hands are oversize and certainly not "realistic," but they hold the saw and board with the kind of convincing authority that would later lure Jacob Lawrence to the hands of manual laborers. Van Gogh and Lawrence continued an even older tradition. Throughout the ages the hand has remained an important subject and symbol in the visual arts, and to this day instructors in art classes tell students to draw their own hands as practice. Salvador Dalí is said to have charged more for portraits if they included hands. From Egyptian tomb paintings to medieval tapestries, stylized hands convey a spectrum of gestures. In Spanish frescoes of the Romanesque period, rows of saints hold their hands together in prayer as if frozen in the act of clapping. The arts record no more "decisive moment"—to borrow a phrase from the photographer Cartier-Bresson—than the one Michelangelo portrays in the *Creation* on the Sistine ceiling: God's hand about to touch Adam's.

Because we have always revered its talents, the hand is one of the first objects of which we find a visual record. For Paleolithic artists, as for Edwin Landseer in his portrait of Paganini, hands were both tool and model. Again and again in primitive art the hand shows up as a decorative motif. It is easy to make an imprint of the hand by merely pressing it into pigment before holding it against a rock. Stencils were fashioned by using a blowpipe to spray pigment over the spread fingers of the hand. Such prints stenciled by the Anasazi decorate the walls of Canyon de Chelly in Arizona. On the other side of the world, in Carnarvon National Park in Queensland, rock surfaces are

decorated with silhouettes of early Australians' two most important implements—boomerangs and hands. Many caves reveal
hand silhouettes with mutilated or completely missing fingers.
Disease, frostbite, and battles with animals each would explain
some of the missing fingers—but not all. In western Spain, at a
cave named Maltravieso, are many silhouettes of hands missing
the last two joints of the little finger. Archaeologists first attributed these variations on a theme to ritual amputation, but ultraviolet light reveals that originally the imprints were of complete
hands. Some fingers and portions of fingers have been painted
out, leading to a more recent theory that such images represent
hand signals known to others in the region. Archaeologists still
conjecture that some of the images around the world do represent accidentally or deliberately abbreviated fingers.

The human hand's great paradox is that it has both carved
the gargoyles atop Notre-Dame and activated the atomic bomb
that fell on Hiroshima. Symbolic hands can represent both op-
pressor and savior. In the nineteenth century, a silhouette of our
primate forepaw was the symbol of the Black Hand criminal society in Sicily. By the late 1960s, a raised fist represented Black
Power. Clasped hands also denote community groups and organi-
zations such as the Rainbow Coalition. In 1977, after the anti-
apartheid activist Stephen Biko was murdered by the South
African Security Police, he was buried in a coffin bearing an even
more powerful symbol—two hands ripping apart the shackles
that had bound them.

FINGER MATH

*And the narrowest hinge in my hand puts to scorn all
machinery. . . .*

WALT WHITMAN

The most striking characteristic of the hand is its five independent digits. Although some possess fewer, no mammal, bird, rep-
tile, or amphibian normally has more than five. This characteristic

is called *pentadactyly*, "five toes." Variations on the theme are considered pathological conditions in humans, and naturally they attract attention. In Toni Morrison's novel *The Bluest Eye*, the young narrator and her best friend mock a popular classmate when they learn "that she had been born with six fingers on each hand and that there was a little bump where each extra one had been removed. . . ." Another such abnormality is *syndactyly*, in which the fingers are joined together by a web of skin. *Ectro-dactyly* refers to the loss of a digit, as in the "thumbless" Colobus monkeys, or the lost index finger in the golden potto. For years scientists thought that the pentadactyl body plan was all but handed down from nature's Mount Sinai, because it is the norm for most nonfishy vertebrates. Since then, however, studies—of the Devonian amphibians *Ichthyostega* and *Acanthostega*, for ex-ample—have revealed that some early tetrapods had more than five toes. It is now known that the quantity of digits is not com-manded as much as nudged by DNA, with the development of the hand requiring specific chemical orders at certain decision points. A number of extant amphibians are only four-toed, and studies have determined that this kind of ectrodactyly results from a fusion of the fourth and fifth digits during individual development.

All developmentally normal mammals have the same num-ber of bones per finger. For that matter, so did some of their an-cestors, the reptiles that evolved into the first mammals. This signature "phalangeal formula" of hand bones for mammals is 2.3.3.3.3, which refers to two bones in the thumb and three in each of the other fingers. The current English names for our own version are, in order, the thumb, index finger or forefinger, mid-dle finger, ring finger, and little finger or pinkie.

The cultural response to our five-fingered mammalian fore-paws is a good example of how far culture can run in unexpected directions while responding to natural phenomena. Once we evolved the idea of numeration, we first counted our own fingers. The word *digit*, which nowadays we use as a noun for numbers and adapt into an adjective root for the computerized trinkets around us, is from the Latin word for the fingers and toes. In contrast with periodic natural occurrences—day, month, year—imaginary milestones such as the decade, century, and millennium are inevitable side effects of the decimal system. As school-children learn to their dismay, there are other systems of numeration (the Sumerians used sexagesimal, and computers use binary) but, because it is rooted in our ten mammalian digits, the decimal system is the most common. Incidentally, the Sumerian base-six system is the origin of our reckoning time in sixty seconds per minute and minutes per hour, and in the measurement of spherical angles as seconds, minutes, and degrees on the surface of the Earth and in the map of the heavens.

Our five digits are fluent in the ancient language of manipulating the world around us. They can work together as a single hand like the paws of our distant forebears, or they can exploit the greater phalangeal individuality that is our more recent primate heritage. Therefore each of our fingers must have nerves and tendons to direct movements that may need to be either individual or united. To bend these digits requires nine flexor tendons in each hand, a pair for every finger and a single for the one-jointed thumb. The Latin word for the wrist is *carpus*, and from it we get the term "carpal tunnel" for the channel in the wrist through which so many nerves and tendons must crowd. Imagine the half-inch-wide median nerve from the spinal cord as an eight-lane highway running down the arm, carrying a flow of traffic that must cross a two-lane bridge before fanning out toward five separate destinations—and this traffic bottleneck is already crowded with all those tendons. The wrist is trouble-prone because it has too much going on in its narrow channel.

Although we do not yet understand all of its causes, one result of this lack of evolutionary foresight is carpal tunnel syndrome, or CTS. "In a medical sense," writes neurosurgeon Frank T. Vertosick, "CTS is a nonpenetrating version of the stigmata." The word *stigmata* derives from the Greek meaning "to puncture with a sharp weapon," but to the Catholic Church it means the wounds suffered by Jesus during his crucifixion—six of them, including the scratches from the crown of thorns but counting them as a single wound. In *The Day Christ Died*, Jim Bishop argues that scenes in painting and film portraying Christ as crucified with nails through his palms are historically inaccurate. Vertosick provides the medical background behind Bishop's thesis. Experiments on cadavers demonstrate, says Vertosick, "that a man crucified in this manner couldn't hang from a cross without the nails tearing out between the middle and ring fingers." The point of crucifixion, which was a common form of execution at the time, was both to inflict prolonged agony upon the offender and to display this suffering as a cautionary advertisement to other wrongdoers. Both greater pain and more reliable hanging of the body were accomplished by driving nails not through the palm but through the wrist. As Vertosick writes, "The heaviest body can hang indefinitely from nails driven through this area, thanks to the great strength of the ligaments binding the carpal bones together." And the pain resulting from piercing the median nerve is excruciating.

We do not have to undergo crucifixion to experience pain in this troublesome region. Carpal tunnel syndrome results from repetitive motion stress in the wrists. "The large number of nerve cells dedicated to hand sensation," notes Vertosick, "also makes us hypersensitive to hand pain in the same way that a heavy nerve supply to the face and head makes us susceptible to headaches and facial tics." Employing a hand that evolved to perform more varied tasks, many contemporary professions involve work that can inflict CTS—from data entry to running a jackhammer, from carpentry to writing books.

One conspicuously trouble-prone group is pianists. The nineteenth-century German composer Robert Schumann is a case in point, and also an example of how complex our psychological relationship with our own dexterous fingers can become. In his compositions, even as early as the Toccata in C Major (op. 7, published in 1834), Schumann forced piano technique to the limits of the hand's capabilities. His troubles arose from fanatical devotion to practice, as recorded in his diaries of the time: "Two hours of finger exercises—ten times the Toccata—six times finger exercises—twenty times the variations alone—and in the evening it just wouldn't work . . . angry about it—really deeply." Schumann's composition has left generations of pianists angry. It is a difficult piece to perform because it requires the right hand to play two lines simultaneously, one with the thumb and index finger, the other with the fourth and fifth fingers.

When traveling without a piano, sometimes Schumann practiced on a noiseless keyboard whose spring-driven keys kept his fingers nimble. The young composer rationalized his emphasis on dexterity with the claim that "greater mechanical power would give me greater control over the [musical] substance." Later he also tried holding his wrists higher, a posture that relaxes the flexor muscles in the forearms and loosens the fingers' play across the keys. Unfortunately, sometimes the resulting fluency is undermined by an accompanying loss of control.

From an early age Schumann experienced pain in his right hand. He tried all sorts of approaches to decrease pain while increasing dexterity; scholars and musicians still argue over the value and danger of these contraptions. The bandmaster Johann Bernard Logier had invented devices he called "chiroplasts," which forced students' fingers into positions with regular spaces between them. Many teachers recommended finger stretchers of one kind or another. In England people called one version "cigar boxes," and in his diary Schumann refers to a *Cigarrenmechanik*.

Psychiatrist Peter Ostwald, in his study *Schumann: The Inner Voices of a Musical Genius*, examines Schumann's hand troubles at length. The composer's pain coincided with his ongoing attempts to secure from his teacher, Friedrich Wieck, permission to marry Wieck's barely teenage daughter, Clara, who was herself a musical prodigy. Ostwald describes the hand pain as "psychophysiological." He adds later, "An analysis of this problem shows that it was closely connected to his conflicts with Wieck, his competitive feelings regarding Clara, and his desire to become a creative rather than a performing artist." There is no question that Schumann was always tangling up his problems and his motivations. He was an alcoholic and a hypochondriac, but scholars such as Kay Redfield Jamison have demonstrated that he also suffered from manic-depressive disorder. (Schumann so clearly recognized his own distinct mood states that he assigned them pseudonyms, Florestan and Eusebius.) The hand pain that caused a stumbling block in his own performance career did indeed occur just as Clara's star was rising. Ostwald may be correct in his confident gloss of these tangled psychological issues, but there is also the possibility of a straightforward repetitive-motion ailment such as carpal tunnel syndrome.

Schumann was not the last musician to experience such problems. Two leading pianists in the second half of the twentieth century, Gary Graffman and Leon Fleisher, both developed problems with the right hand that became severe enough to prohibit concert performance. Rather than retiring, both turned to the surprisingly large repertoire of music composed for left-handed emphasis. Paul Wittgenstein, brother of the philosopher, became the best-known composer for the left hand after he lost his right arm in World War I. Thanks to Western music's arbitrary emphasis on the right hand, and to a concert pianist's need to broadcast the melody through a concert hall, damage to the left is far less common than the right-hand troubles that plagued Schumann and his musical descendants.

Because impairment of the hand has always endangered an individual's welfare, legal systems over the centuries codified the compensation a victim might expect for the loss of a digit. Most were more detailed than the vengeful litany in the Book of Exodus: "Eye for eye, tooth for tooth, hand for hand, foot for foot. . . ." Alfred, the ninth-century Saxon king of Wessex, England, recorded compensations, and therefore we know his names for the fingers. The thumb was called *duma*, and in order the others were the *scythe finger, midlestafinger, goldfinger*, and *lytlafinger*. Our venerable word *finger* dates from Old English, with cognates in the Danish *vinger* and the Old Norse *fingr*.

Canute, or Knut, the Danish king who invaded England in the eleventh century, created his own system of reparation for lost digits. In his Medieval Latin, the terms were *pollex, demonstratorius, impudicus, annularis*, and *auricularis*. These names deserve a closer look. *Pollex* is still a technical term for the thumb and is the root of the ancient Latin phrase *pollice verso*, the "thumbs-down" signal that sent gladiators to their deaths. *Demonstratorius* seems to allude to the role of the finger as pointer. Conjectures about *impudicus* range from the finger's possible role in insults to its useful length in exploring the female genitals—a theory that assumes that, like *impudent, impudicus* derives from *pudendum*. Because "giving the finger" in a popular rude gesture is not a recent invention, this digit has also been called *obscenus. Annularis* refers to the same object that often decorates this particular finger today—a ring. *Auricularis* comes from the auricle, the outer ear, because the little finger is the one commonly used for cleaning wax from it. Incidentally, the first name of the villain in the James Bond movie *Goldfinger*—who shows up in Alfred's list—was Auric. This was not, however, a reference to the wax-removing abilities of his finger; in chemistry *auric* is an adjective referring to gold.

Another social consequence of the loss of digits occurred when lepers were ostracized. Medicine has lessened the incidence and ferocity of leprosy worldwide, and more sophisticated cultures have largely abandoned their traditional responses, but throughout most of history lepers were treated as pariahs. Among the visual warning signs of leprosy was a missing fingertip or entire digit. Many people had the mistaken notion that such problems resulted from the disease itself, but actually injury occurred when their numbed warning systems failed to serve properly. Loss of sensation is one of the side effects of this dreaded malady.

Working together unimpaired, our fingers can form a cup, a hook, or a fist. They can count or play the piano. Booker T. Washington, the African American educator, aptly demonstrated the mix of cooperation and independence typical of the fingers when he used them to symbolize a segregated but not antagonistic society: "In all things that are purely social we [black and white] can be as separate as the fingers, yet one as the hand in all things essential to mutual progress." The imagery in Washington's optimistic error works both ways. The tragedy of institutionalized segregation in the United States reminds us that once we lose any finger, we no longer have as powerful a hand.

RULE OF THUMB

For all the fingers' collective virtues, one digit literally stands out—Alfred's *duma*, Canute's *pollex*, the resourceful adaptation that Albinus, the German anatomist, called "the lesser hand." Isaac Newton proclaimed that it alone was enough to prove the existence of God. "The hand denotes the superior animal," wrote the nineteenth-century palmist d'Arpentigny, "but it is the thumb which individualises the Man."

The thumb is a primate innovation. Without it, John Napier has written, "the hand is put back sixty million years in evolutionary terms to a stage when the thumb had no independent movement and was just another digit." Paleontologists date the evolution of the thumb's ability to rotate independently, the predecessor to the opposition of which we are so proud, as far back as 18 million years ago. The truly human thumb came much later. It was Napier, an anatomist and paleontologist working with Louis Leakey and Philip Tobias, who determined that the skeletal remains of an ancient hominid, found in Olduvai Gorge in Tanganyika in the early 1960s, revealed evidence of an opposable thumb. Because of its large brain, the species is considered a possible missing link between the early australopithecines and the later *Homo erectus*. But it was the opposable thumb that lent it the ability to chip stone tools, a local cottage industry from which it received its name, *Homo habilis*, "handy man." Volcanic rock around the remains dates between 2 and 1.5 million years ago.

In general, among other mammals, the thumb's equivalent possesses no outstanding virtues. Because it is at the end of the five digits, it has slightly greater mobility, but so does the little finger, which occupies a similar position on the other side of the hand. When we look at primates, however, we find a different story entirely. Beginning with the aye-ayes, lemurs, tarsiers, and other members of the lower primates, proceeding through the Old and New World monkeys and great apes that constitute the higher primates, and culminating (as we love to do) with ourselves, we find a measurable increase in the variety and precision of the thumb's movements.

By comparison with their less accomplished cousins, the other higher primates approach our own level of dexterity. However, although the difference is obviously of degree rather than kind, a single example will illustrate how far apart those degrees fall. In the early 1990s, an Indiana University scientist named Nicholas Toth employed his own clever brain and thumbs in an attempt to teach a bonobo named Kanzi how to make tools out of stone. This is the same ingenious ape whose language aptitude renewed interest in such studies in the 1990s. Still, although he could pick up a stone and pound it against another, Kanzi could not grasp it precisely enough to align the two accurately—and despite his courteous imitation of Toth's actions, apparently he could not grasp the concept of tool manufacture, either. This kind of action is peculiar to human beings.

For our splendid proficiency we can thank the saddle joint of the thumb's metacarpal. "When the joint is flexed," as *Gray's Anatomy* describes it, "the metacarpal bone is brought in front of the palm and the thumb is gradually turned to the fingers. It is by this peculiar movement that the tip of the thumb is opposed to the other digits." Indeed, in this peculiar movement lies a great deal of history. Opposition is a wonderful adaptation.

Occasionally an animal has evolved a makeshift version. This is the case with the "thumb" of the giant panda, made famous by Stephen Jay Gould as an example of the make-do ways of nature. One of the panda's wrist bones, the radial sesamoid, has evolved an imperfect opposition that allows its distinctly unhandy paw to grasp bamboo stalks—the stripping of which occupies the majority of the animal's waking hours. Some African monkeys, such as the spider monkey, have thumbs that are reduced or entirely lost, and are actually called "thumbless" monkeys. In contrast with their Old World cousins, New World monkeys have evolved prehensile tails so versatile that they have been called a third hand. They aid in climbing and even in gathering food. The prehensile tail is so thumblike that some even have a patch of sensitive hairless skin underneath the tip.

With such crucial importance in evolution and history, the thumb shows up prominently in figures of speech. Because this digit does its job only in concert with the other fingers, to be all thumbs means to be clumsy or awkward. Most of us proceed through life by rule of thumb, and—thanks to loss of dexterity—nothing sticks out like a sore one. As the Rolling Stones' song indicates, to be under others' thumbs is to be subject to their control.

The thumb is featured in a number of insults. To thumb your nose at another means to place your thumb alongside the nose and extend or even wiggle the fingers in a surprisingly widespread gesture of derision. A similar affront appears in *Romeo and Juliet*. Abraham, one of the Montague servants, spies Sampson, a Capulet servant, and demands, "Do you bite your thumb at us, sir?"

"I do bite my thumb, sir," replies Sampson.

Abraham repeats his question: "Do you bite your thumb at *us*, sir?"

"Is the law of our side if I say ay?"

"No," says Abraham's companion.

"No, sir," answers Sampson wisely, "I do not bite my thumb at you, sir." Then he cannot resist adding, "But I bite my thumb, sir."

During the Great Depression, hitchhikers were so familiar a sight in the United States that they acquired the nickname "thumbers." To ride your thumb meant to use the characteristic thumb gesture to hitch a free ride. In Tom Robbins's 1976 novel *Even Cowgirls Get the Blues,* Sissy Hankshaw is born with gigantic thumbs. (Uma Thurman was outfitted with oversize digits for the movie version.) Young Sissy looks up *thumb* in the dictionary: "the short, thick first or most preaxial digit of the human hand, differing from the other fingers by having two phalanges and greater freedom of movement." She adopts the last phrase as her mantra, accepts Freud's claim that anatomy is destiny, and becomes the world's greatest hitchhiker.

In the arenas of imperial Rome, gladiators—originally mostly slaves, criminals, and prisoners of war, but later including some professionals—filed past the royal box with cries of "Hail, Emperor, those who are about to die salute you!" While a band played and peddlers hawked snacks, the gladiators fought against animals and against each other, usually in grotesquely mismatched battles. Triumph and extraordinary courage were much admired and could even win freedom. But poor fighting or cowardice, or merely a fall, could result in cries from the crowd of "Cut his throat!" At such a moment, everyone turned to the emperor to watch for a gesture clearly visible from a distance—*pollice verso,* "thumbs down," which meant death to the fallen man. Sometimes crowd members indicated their preference by the same gesture. There was no thumbs-up for those to be spared. If the crowd or the ruler wanted to prevent a man's death, they held out fists with the thumbs clenched inside.

Both *pollice verso* and the ahistorical thumbs-up persisted in the movie ratings of Roger Ebert and Gene Siskel. Frequently they disagreed, resulting in the confusing signal of one thumb up and one down. Unlike, say, Trajan's, Siskel's and Ebert's thumbs didn't signify life or death, but they could affect a movie's attendance and therefore its earnings. A magazine profile about them once showed their critical digits drawn like swords, accompanied by the headline WHOSE IS BIGGER? In 1995 the esteemed Looney Tunes director Chuck Jones released a new short animated film, *Another Froggy Evening*. In this adventure Michigan J. Frog finds himself in ancient Rome, and the critics in the stands (including a caricature of Chuck Jones) express their opinion of the amphibian's performance by turning their thumbs downward. Only two members of the audience hold their thumbs aloft—animated versions of Siskel and Ebert.

Over the centuries the prominence, importance, and sensitivity of this magnificent adaptation have inspired the small but inventive percentage of humans who become torturers. Thumbscrews were instruments of torture in which the thumb could be squeezed or compressed. When not killing their enemies, soldiers have often removed their thumbs—thus depriving them of the ability to hold a weapon while leaving them alive as cautionary lessons. In the Bible, Judah and Simeon his brother go after the Canaanites and the Perizzites, and they find the enemy king, Adonibezek, in the city of Bezek: "But Adonibezek fled; and they pursued after him, and caught him, and cut off his thumbs and his great toes." The additional injury and insult of slicing off the toes deprived the enemy of both dexterity and balance. In Michael Ondaatje's novel *The English Patient*, and in the film version, Nazi torturers cut off a character's thumbs when he refuses to provide the information they want. Lacking opposition, the man's hands are reduced almost to flippers. It is a striking reminder of the crucial role played by the lesser hand.

OUR NATAL AUTOGRAPH

We appoint snowflakes the very emblem of uniqueness because no two are alike. Yet no two objects of any kind are truly identi-

cal. Perhaps a more instructive example of the snowflake principle is one we carry with us. Like a cautious bureaucrat, nature refuses to assign two human beings the same fingerprints. Not that
fingerprints are the only reliable method of identifying someone.
Because similarly patterned areas of the foot also are unique and
unchangeable, hospitals routinely footprint newborn infants. As
a means of identification, however, the patterns on fingertips
have two advantages: Fingerprinting is a simple procedure, and
criminals seldom ransack apartments or fire handguns with their
feet.

The moment that fingerprints were established as *the* form of
identification, they became a burden for those wishing to avoid
notice. John Dillinger is perhaps the most famous of the numerous miscreants who have tried to obliterate their fingerprints. He
underwent painful operations to remove his prints—all to no
avail. The patterns grew back each time, just as identifiable as before. Dillinger had encountered one of the more arcane talents of
the human body. We marvel at a lobster's ability to regenerate a
lost limb, but fingerprints top even this accomplishment. Each
new lobster claw is smaller than its predecessor. However, unless
the deeper layers of the dermis are seriously damaged, when the
skin of a fingertip is injured the body remembers its own blue-
print and re-creates the pattern. The skin grows back every whorl,
every spur and loop and arch, like a scarred Rembrandt magically
repairing itself.

An infallible method of identification has virtues beyond that
of tracking criminals. Accidents, natural disasters, war, terrorism—
all leave in their wake victims who require identification. Most
of the fingerprints on file are not those of criminals. Both law-
enforcement and civic agencies advise that everyone (at least
everyone not contemplating a life of crime) be fingerprinted. As
a result of their importance, finger patterns have a technical
name, *dermatoglyphics*. Sometimes the same term is applied to the
study of them, but for that we also have the word *dactylography*.
The terminology currently in use indicates the degree of precision that dactylography has attained. There are plain and tented

arches, ulnar and radial loops, and a variety of whorls, including plain and accidental. Within these patterns we find further variety. Ridges bifurcate and dead-end. They form bridges, spurs, and loops. There may be short ridges that don't connect to any others, or even tiny ones called islands. These variations are classified by a formula expressed in both letters and numbers, including primary, secondary, small-letter, subsecondary, major, final, and key. We possess these detailed forms of identification long before a nurse presses our inked hands and feet to a birth certificate. Around the end of the second month after conception, about the time the embryo begins to look less like a Dr. Seuss character and more like a human being, the fingers and toes are clearly recognizable as such. Within the next month or so, patterned ridges begin to appear on them.

Like every other aspect of the human body, fingerprints have a natural explanation. Form does indeed follow function, and distinctive marks on the fingers are not merely nature's label certifying that we are all homemade. They have a purpose. Our versatile skin—flexible around the knees and elbows, taut across the palm, varying its thickness from the astonishingly thin eyelid to the callused sole of the foot—has many jobs to do. Not least is its role in gripping. With a magnifying glass, you can see that the patterns on the fleshy pads of your fingertips consist of ridges of epidermis with little valleys between them. These ridges are similar to a tire's tread, but of course they are far more versatile and responsive. When you lift a wineglass, you employ your independent fingers and opposable thumb, but still you might drop it if not for the plierslike grip of the skin on your fingertips. The ridges result from a thickening of the epidermis below and above the skin's surface, which not only aids in grasping but also houses the openings of the sweat glands. They are called *papillary ridges,* or *dermal papillae,* from the Latin term for nipple, *papilla,* which biologists employ for any such small protuberance. The combination of the ridges and the sweat glands decrees that human beings leave oily fingerprints on everything we touch. But it is their random growth patterns that led to fingerprints' historical significance.

@

Although many people through the ages must have noticed their finger patterns, full appreciation had to await magnification. In the 1600s the telescope and the microscope began to reveal the limitations of human perception and to hint at how technology might expand it. During this remarkable century, three men in three different countries contributed to our growing awareness of fingerprints. The first was an Italian biologist and anatomist, Marcello Malpighi, the man who discovered capillaries and thereby supplied the missing link in William Harvey's explanation of blood circulation. Malpighi examined the tips of fingers—along with everything else that came his way—under a magnifying glass. He identified the dermal papillae as minute ridges that contain capillaries and even nerve endings. In 1684 an English physician and botanist with the Dickensian name of Nehemiah Grew realized that these ridges form designs "very regularly disposed into Spherical Triangles and Ellipticks." Sweat also was clearly visible, so that "every Pore looks like a little Fountain." The next year the Dutch anatomist Govard Bidloo published his *Anatomia Humani Corporis*. Among the beautiful drawings of microscopic close-ups of the skin is a thumb featuring a simplified but clear fingerprint.

Later developments were equally international. In 1823 the Bohemian physiologist Jan Evangelista Purkyne published a description of nine kinds of patterns he had found in fingerprints. His categories basically agree with those used nowadays, and he is credited with the first suggestion that fingerprints could be classified by pattern. The next historic step occurred when Henry Faulds, a Scottish medical missionary in Japan, wrote a letter to the scientific journal *Nature*. It was published on 28 October 1880, under the title "On the Skin-Furrows of the Hand."

> A large number of nature-prints have been taken by me from the fingers of people in Japan, and I am at present collecting others from different nationalities,

which I hope may aid students of ethnology in classification. . . .

When bloody finger-marks or impressions on clay, glass, &c., exist, they may lead to the scientific identification of criminals. . . . There can be no doubt as to the advantage of having, besides their photographs, a nature-copy of the for-ever-unchangeable finger-furrows of important criminals. . . .

Apparently *Nature* quickly made its way throughout the still-sunlit empire. Barely a month later, there appeared in its pages a reply to Mr. Faulds from the government administrator of the Hooghly district in Bengal, Sir William Herschel, who was descended from the eminent English astronomer of the same name. He seemed eager to establish his own priority in the matter of fingerprints.

I have been taking sign-manuals by means of finger-marks for now more than twenty years, and have introduced them for practical purposes in several ways in India with marked benefit.

The object has been to make all attempts at personation, or at repudiation of signatures, quite hopeless wherever this method is available. . . .

Faulds and Herschel were spearheading a revolution. Eventually it would overthrow the most popular method of identification in use at the time, anthropometry, a system formulated largely by France's Alphonse Bertillon. Anthropometry measured the proportions of body parts—the length of the middle finger, the breadth and height of the head, the length of the forearm and foot, and so on. When possible, photographs were made. Although the measurements added up to 243 subdivisions, in retrospect the system's shortcomings are obvious.

Francis Galton, the Victorian scientist generally considered the founder of eugenics (the movement to improve humanity through selective breeding), took the work of Faulds and Herschel a step further. In 1892 he published *Finger Prints*, in which he proposed the method of fingerprinting still in use today. First a thin layer of ink is spread on a metal or glass plate. To record each digit's full pattern, the finger is rolled across the ink and then rolled on a white card. Galton also proposed collection methods and devised a complicated system of classification. However, a committee that was overseeing the introduction into Britain of the anthropometric method found the complexities of Galton's system daunting and recommended that it be used only as a supplement to Bertillon's. The disrespect leveled at dermatographers in Galton's era is apparent in his defensive tone. "Let no one despise the ridges," he said of fingerprints, "for they are in some respects the most important of all anthropological data." What truly deserved scorn were the biases that Galton claimed his dermatoglyphics confirmed. For example, he imagined that the fingerprints of people of African descent revealed "greater simplicity" than did those of Europeans—and even betrayed "the general clumsiness of their fingers."

The first truly practical method of fingerprint identification was invented in Argentina, by an expatriate Dalmatian named Juan Vucetich, who ran the Anthropometric Bureau of Identification in Buenos Aires. He elaborated upon Galton's system, making it more complete and reliable. In 1891 his became the first method to be officially implemented, and it is still in use in some South American nations. Shortly afterward, Edward Henry introduced a somewhat different system in England, which was soon in use at Scotland Yard and eventually in most of Europe and the United States. Years later Vucetich told a conference that "never, in spite of all our efforts, were we able to determine with certainty the identity of an individual by means of measurement because we always found differences for the same person. For this reason we adopted fingerprints."

This famously simple method of identification also enables

investigators to quickly fingerprint corpses, because fingerprints persist longer after death than any other identifying feature except bones and teeth. Even cadavers that have undergone considerable decomposition can still provide fingerprints. After death, as the underlying tissues decompose, sometimes the skin of the hand can be removed like a glove—minus the nails, which usually fall off. Then, in a grotesque impersonation of identity, a laboratory technician dons the removed hand skin, inks the deceased's fingertips, and rolls the fingers on white paper.

Sometimes fingerprints even outlast the memory of a crime. In 1952 peat cutters found a man's body in a bog near the village of Grauballe in Denmark. His throat had been slit. At first the workers assumed that they had found a recent murder victim, but it turned out that the corpse dated from the early Iron Age. The body, like that of the famous Lindow Woman and Tollund Man and others, had been kept intact by humic acid produced when the bog plant sphagnum decays. By killing bacteria, humic acid forestalls putrefaction. The hands of Grauballe Man were so well preserved that his fingerprints were still intact—more than two thousand years after his death.

When Galton's *Finger Prints* came out in 1892, one of his readers was an American novelist who claimed that he eagerly "devoured" the book. As a consequence the next prominent appearance of fingerprints, two years later, was not in a dry treatise but in a satire of racial attitudes in the Old South—*The Tragedy of Pudd'nhead Wilson*, by Mark Twain. Twain kept an eye on advancements in science and technology; in fact, he was the first prominent novelist to write a book on the newfangled typewriter. After incorporating Galton's ideas into his new novel, Twain wrote to his publisher that "the finger-prints in this one is virgin ground, absolutely fresh, and mighty curious and interesting to everybody." Later he wrote, "The fingermark system of identification . . . has been quite thoroughly & scientifically examined by Mr. Galt [*sic*], & I kept myself within the bounds of his ascertained facts." Marginal illustrations in the first edition of the book include actual fingerprints.

Twain's detective, David "Pudd'nhead" Wilson, is a lawyer in Dawson's Landing, Missouri, a man with a talent for observation. In his methods he is reminiscent of both Sherlock Holmes and Voltaire's Zadig. But he solves the case in court, à la Perry Mason, by identifying the bloody fingerprints upon the murder weapon. Wilson reminds the court that for twenty years his hobby has been collecting and cataloging the fingerprints of local citizens. With his collection he identifies the true murderer.

As Wilson explains the uniqueness of what he calls our "natal autograph," the court officers and spectators discover their own fingerprints.

> Every man in the room had his hand up to the light, now, and his head canted to one side, and was minutely scrutinizing the balls of his fingers; there were whispered ejaculations of "Why, it's so—I never noticed that before!"

TOBIN'S PALM

Fingerprint analysts are not the only ones who "read" the body. So do physicians, detectives, psychologists, and anyone who interprets facial expressions or body language. But not every interpreter has equally rigorous standards. While fingerprints provide a practical method of identification, the lines in the palm have inspired more fanciful responses. The difference lies in imagining connections between unrelated phenomena: linking creases in the skin with an individual's destiny.

In O. Henry's story "Tobin's Palm," the narrator accompanies his friend Tobin to visit Madame Zozo, the Egyptian Palmist. She examines "the line of his fate" that runs across his palm. Immediately she perceives trouble with a sweetheart, but she divines the woman's initials only after overhearing Tobin whisper her name. Madame Zozo warns him to watch out for a dark man and a light

woman, predicts a financial loss and a voyage upon water, and assures Tobin that a crooked-nosed man with a long name contain-
ing the letter *o* will bring him good fortune. All of Madame
Zozo's predictions come true on the same day—except that the
name of the crooked-nosed gentleman, Maximus G. Frieden-
hausman, lacks the letter *o*.

It is easy to sympathize with Tobin's desire to know the future. With our fellow creatures playing, laughing, making tools,
disseminating technological advances, and banding together to
kill their own kind—activities that until recently we proudly
claimed as solely human—anxiety is one of the few traits that actually may be unique to our species. Thanks to this apprehension
and to reason's meager role as a force in civilization, fortune-
telling is an ancient practice. Most religions feature apocalyptic
prophecies of one kind or another, but many believers seek out
forecasts with more immediate personal relevance. Over the millennia we have tried to discern character and predict future
events from the bumps on the head, the shape of the nose, the
viscera of sacrificial animals, the position of the sun against the
stars, and even the flight of birds. Inevitably, somewhere along
the way imaginative souls invested their own skin with significance. Palmistry has been around for many centuries. Obscure
remarks that may refer to the belief appear in two books of the
Bible, Proverbs and Job, and at least passing references survive in
the works of Aristotle and Pliny the Elder. Of course, then as
now, skeptics countered believers. Juvenal, the Roman satiric
poet, mocked "the cheap chiromancer's art." The word *chiromancy* derives its name from two Greek words meaning "divining
by hands."

Chiromancers assign meaning to almost every topographical
feature of the palm. In case you apply the following description
to your own hand, note that cupping the palm slightly makes the
lines more prominent, and even then some remain indistinct.
The crease that begins closest to the center of the palm and
moves outward to a point between the index finger and the

thumb is called the "head line." The "heart line" is the crease that runs somewhat parallel with the head line but closer to the fingers. It begins under the first or second finger and curves to the edge of the hand below the little finger. The "line of life" basically outlines the Mount of Venus—not to be confused with a similarly named region elsewhere on many bodies—at the base of the thumb. The crease that Madame Zozo examined on Tobin's palm, the "line of fate," runs down the middle of the palm and continues toward the wrist. Many less prominent lines have also been given names and meanings.

In order, from the index finger to the little finger, the digits are the Finger of Jupiter, of Saturn, of Apollo, and of Mercury. Opposite the Mount of Venus, across the palm, lies the Mount of the Moon. These names for parts of the hand demonstrate palmistry's links with another venerable superstition, astrology. To anatomists these mounds are, respectively, the *thenar* and *hypothenar eminences;* they are simply pads of fatty tissue. There are a number of further classifications. In some systems the hand is divided into zones called active, passive, inner, and outer. Some palmists claim that the "major" hand, which is the one that we favor, reveals our experiences in the outside world, while the "minor" hand documents inner emotions and traumas. Osbert Sitwell co-opted chiromantic metaphors when he titled his autobiography *Left Hand, Right Hand!* because, "according to the palmists, the lines of the left hand are incised inalterably at birth, while those of the right hand are modified by our actions and environment, and the life we lead." Many palmists still employ a system begun in 1839, when a retired French soldier named Casimir Stanislas d'Arpentigny described six types of hand—elementary, spatulate, psychic, square, knotty or phylosophic, and conic. Later he added a useful seventh category, "mixed." A number of palmists insist that the lines of the foot are equally viable sources for reading character and fate.

Of the lines in the hand, twentieth-century palmist Marcel Broekman wrote, "Their cause is unknown. But they speak very clearly, in a complex language." Actually they speak very clearly in a simple language; the origin of the lines in the palm is quite well understood. How tightly the skin of the hand is attached varies. On the back, for example, it is extremely loose to facilitate movement. The skin of the palm is some of the least flexible on the body, as you quickly learn if you try to pinch it between your fingers. It is this tautness that assists the fingers in gripping and permits such actions as turning a doorknob. Firm anchoring to the tissues around the joints results in creases called *flexure lines*. They are not merely wrinkles but permanent "hinges" that indicate where the skin is bound to the underlying tissues. Like the folds in clothing, flexure lines do not precisely indicate where the joint lies. Between the meeting of bones that creates a joint and the skin above lie hidden tissues and tendons that, like geological strata, influence the location of surface features. Still, although flexure lines cannot predict the future, they roughly outline where the hand can be expected to move, and thus hint at its extraordinary facility.

A palmist who examined the hand of one of our primate cousins might find the animal's future difficult to predict. In non-human primates the head and heart lines form a single fold called the *simian crease*. This pattern is visible from lemurs to tarsiers, from the macaque to the gorilla. Palm readers may occasionally encounter this phenomenon without going to the zoo, because it occurs in about half of 1 percent of human beings. Humans usually lack a simian crease because the different wrinkles reflect the different adaptations of the hand. Nonhuman primates have an index finger with little independent movement; it works in concert with the other fingers. We, in contrast, have highly flexible and adaptable index fingers, and the creases in the palm reflect this uniqueness.

One recent history of palmistry includes a photograph of the palm of a chimpanzee for cross-species comparison. Apparently it didn't occur to the author that such an evolutionary similarity might invalidate the mystical significance of the lines in the

human hand. Or perhaps the fate of chimpanzees is simply waiting for Madame Zozo to decipher it.

ON THE OTHER HAND

The letter in the newspaper bore the title "A Petition to Those
Who Have the Superintendency of Education." It began humbly:

> I address myself to all the friends of youth, and
> conjure them to direct their compassionate regard to
> my unhappy fate, in order to remove the prejudices of
> which I am the victim. There are twin sisters of us;
> and the eyes of man do not more resemble, nor are
> capable of being on better terms with each other than
> my sister and myself, were it not for the partiality of
> our parents, who made the most injurious distinction
> between us. . . . She had masters to teach her writing,
> drawing, music and other accomplishments, but if by
> chance I touched a pencil, a pen, or a needle I was
> bitterly rebuked; and more than once I have been
> beaten for being awkward, and wanting a graceful
> manner. . . . Condescend, sirs, to make my parents
> sensible of the injustice of an exclusive tenderness,
> and of the necessity of distributing their care and
> affection among all their children equally.
>
> I am, with profound respect, Sirs,
>   Your obedient servant,
>   The Left Hand

   The author of the letter was Benjamin Franklin. No doubt
his subscribers chuckled over Franklin's successfully pulling their
leg once again. But he made his point. Seldom has the prejudice
against left-handedness been better presented. Of course,
Franklin did not write the letter merely out of empathy; he was
himself left-handed.

Franklin was addressing an age-old conflict. "In the use of the hand," declared Plato, "we are, as it were, maimed by the folly of our nurses and mothers, for although our several limbs are by nature balanced, we create a difference in them by a bad habit." He insisted that we ought to enforce our natural ambidexterity and train ourselves and our children to use both hands. Usually, disagreement with Plato is a matter of opinion, but on this point he simply erred. Other than in their bilateral symmetry, our limbs are not by nature balanced. An inborn predisposition toward right-handedness is a trait shared by the great majority of human beings throughout history, but with a consistent proportion of the populace born left-handed instead. Estimates vary, but the general consensus is that roughly 10 percent of the population is left-handed.

Variety alone should prevent generalizations about left-handers. Queen Victoria, Mark Twain, the Boston Strangler, and Marilyn Monroe were all left-handed. The college papers of a future emperor of France were almost illegible because it was awkward for the young Napoleon to write from left to right. The pattern of wounds on some of Jack the Ripper's victims led investigators to conclude that he wielded his surgically precise knife with his left hand. Leonardo da Vinci drew sequential drawings, as in his studies of the foreleg of a horse, from right to left—and, unlike most right-handed artists, drew profiles facing right.

Yet bias against the minority has plagued left-handers throughout history. In many places around the world, the left hand, even the entire left side of the body, was considered inferior, if not evil. There was even a medieval belief, fully worthy of Monty Python, that infants destined to become saints gave early hints of their preternaturally acute piety by refusing to suckle at their mother's left breast. Supposedly the devil lurked behind the left shoulder—which is why, when you spill salt, you should throw some of it over your left shoulder and into the devil's face. Many cultures employ what anthropologists call "dual symbolic classification," in which dichotomies such as right/left are linked with male/female or solar/lunar or especially good/bad. "The world is

divided into pairs of opposites," writes psychologist Chris Mc-
Manus about this kind of thinking, "which are then tied together,
so that given one member, the others are also known." Politicians
wage entire campaigns with such simplistic formulations.

Many Muslims still prefer to eat with the right hand, because
of the ancient associations of the left with hygiene, and some
Christians insist upon taking communion with the right. The majority of Indians touch food only with their right hand. Jesus
claims that God will sort humanity into two groups on Judgment
Day, gathering the righteous on his right: "Then shall he say also
unto them on the left hand, Depart from me, ye cursed, into ever-
lasting fire." To this day many churches are designed with left and
right firmly in mind. Michelangelo is said to have been left-
handed, or at least ambidextrous. While standing atop the elaborate scaffolding below the Sistine ceiling, il Divino painted—
supposedly with first one hand and then the other—Jesus' vision
of sorting souls by hand. God points to heaven with his right
hand, and to hell with his left. Elsewhere on the same ceiling, in
the Creation of Adam, God and the first man reach toward each
other. God is about to bestow life with, of course, his right hand.
Yet Adam appears to be a southpaw. Languidly, with his arm resting on his upraised knee, he stretches out the forefinger of his left
hand. Perhaps, as Michelangelo sketched the scene, composition
simply trumped symbolism.

Nowadays, even though the prejudice is fading, most items
from bowling balls to watch stems are still designed for the right-
handed. In the United States, the only widespread machines that
are not intended for lefties but are easily accessible to them are
highway tollbooths and the drive-through windows of banks and
fast-food joints. There are exceptions. Remington manufactures
left-handed shotguns and rifles; Yashica produces a left-handed
camera; Timex manufactures a watch for left-handers.

Etymology preserves the history of bias against lefties. Dextrous
and adroit come from words referring to the right hand, while

sinister, awkward, and *gauche* relate to the left. Indeed, the adjective *sinistral* means "left-handed" or "coiling to the left" (counterclockwise). The very word *ambidextrous,* which we now use to mean equally proficient with either hand, actually means "two right hands." We might naïvely expect a more rational attitude from the sciences, but many of the supposedly scientific pronouncements about handedness have been equally cruel. The nineteenth-century Italian criminologist Cesare Lombroso, whose work shows up so often in this book, claimed that left-handedness, along with other traits such as protruding ears and narrow foreheads, indicated an inborn predisposition toward degenerate behavior. As late as the 1940s, psychiatrist Abram Blau wrote that left-handedness "is nothing more than an expression of an infantile negativism, and falls into the same category as . . . a general perverseness." One theory attributed the "abnormality" to hormonal imbalances experienced by the subject's mother during pregnancy.

The Zuñi of the American Southwest may have intuited a deeper understanding of handedness. In their mythology the right side of the body represented impetuosity and restlessness, while the left symbolized reflection and enlightenment. Thomas Huxley could not resist commenting upon such prescientific acuity: "Ancient traditions, when tested by the severe processes of modern investigation, commonly enough fade away into mere dreams; but it is singular how often the dream turns out to have been a half-waking one, presaging a reality." The Zuñi's dream of a mental dichotomy between the mirror-image sides of the body has proven insightful. Our two sides, as practically everyone has heard nowadays, are controlled by different halves of the brain. In most human beings, the left hemisphere is the more analytical, processing information in logical, sequential, mathematical ways. It dominates in language use, from hearing and speaking to reading and writing. The right hemisphere sees the world more holistically; that is, it recognizes patterns—the overall appearance of a face, the structure of a melody, the composition of a work of art. It does not rely upon a reductionist analysis of components to determine a whole. The right hand is controlled by the left hemisphere and the left hand by the right. In his popular work *Drawing Hands,* the Dutch artist M. C. Escher portrayed the interdependence of the left and right sides of the body: a left hand sketches a right hand, which in turn draws the left.

Theories about the origin of handedness range from Plato's foolish nurses to the claim that newborn infants cry less when held near their mother's heart. It is true that, like our closer primate cousins, we generally carry our young on the left side. There are also several ways in which the unborn and the newborn favor their right sides. Ultrasound explorations demonstrate that a fetus sucks its right thumb far more than its left, beginning as early as thirteen weeks into pregnancy. Although test results to confirm it are uncertain, there is also a theory of handedness based upon the relative freedom of the right hand during crucial prenatal development. For the last several weeks before birth, roughly 75 percent of fetuses hold the right arm out toward the mother's abdominal wall, while the left arm is confined against the mother's vertebrae.

Like Lombroso and others, some scientists still regard left-handedness as the result of a problem in the body. However, the pathological explanation does not accommodate all the evidence any more than does the environmental school of thinking. Like so many other aspects of the human body, handedness cannot be attributed to a single cause. A small percentage may trace their handedness to some kind of malfunction in the womb or during birth, but it seems that most lefties inherit their sinistral bent. The number is about two points higher for women and also varies slightly among cultures around the world, but the percentage of human beings who are distinctly right-handed is about 85 to 90. There is no question that handedness involves some kind of genetic component. If you are left-handed, the odds of your child being left-handed are much higher than otherwise, yet still only about fifty-fifty. The mother's influence, either genetic or epigenetic, seems stronger than the father's; left-handed mothers have left-handed children more often than do left-handed fathers.

Why such an adaptation is not found throughout the population is the intriguing question. Genetic mapping and studies of the physical asymmetry of the brain are trying to answer it.

Nor are we alone in our handedness. This common misconception has joined the ever growing list of ways in which we are not unique despite our claims of elite status in the animal world. In 1905 an English social activist named John Jackson founded the Ambidextral Culture Society. "There is no disadvantage, but every advantage, in our being truly ambidextrous!" he wrote. He wanted humanity to return to the sort of natural ambidexterity that he had observed in apes in zoos. "Why should not perfect Ambidexterity be possible now? WHY CANNOT MAN BE AMBIDEXTROUS AGAIN?"

Jackson was as mistaken about animals' ambidexterity as Plato was about humans'. Perhaps his observations at the zoo were not as meticulous as they might have been. By the 1940s scientists were testing the handedness of captive chimpanzees, and later tests were performed on their wild relatives. Gorillas divide down the middle—half right-handers, half left-handers. Chimpanzees seem to slightly favor the right hand. Many other creatures demonstrate a preference for right or left. Fish tend to veer one way rather than the other. Some amphibians always use the same foot when testing or moving an object. In his 2002 book about handedness throughout biology and physics, *Right Hand Left Hand*, Chris McManus tells the story of the twelfth-century Chinese emperor Hui Tsung. When artists completed a painting of the imperial peacocks in their garden, Hui Tsung pointed out an error. One peacock was lifting its right foot to step into a flower bed. "A peacock," noted the emperor, "always raises its left foot first to climb."

It is important to emphasize how basic—in the very structure of the universe—handedness turns out to be. A useful reminder that it is a natural phenomenon was provided by the first official "Siamese" twins. Identical twins grow from an incompletely divided ovum, and *enantiomorphism* (mirror imagery) is the rule in such cases. When Chang and Eng were born in 1811—in Siam, but to Chinese parents—they shared a trait now known to be common to such pairs. Joined at the waist, the twins were exact enantiomorphs. The hair on their scalps grew in opposite directions. The fingerprints on Chang's left hand more closely matched those on his brother's right than on his own. And one was right-handed while the other was left-. Just as Sophocles uttered the artistic version of Freud's dogmatic metaphors, so Lewis Carroll seems to have intuited the paradoxes of modern biochemistry and physics. It is a Looking-Glass universe.

Even more interesting than the weird ramifications of handedness in living creatures is its existence throughout the inanimate world. Asymmetry is the rule of nature. Most climbing plants that twine around a branch or another plant do so by helically twisting to the right, but others are left-turning. Although called "handedness" because it was first observed in human beings, this right or left orientation shows up in many aspects of the natural world, all the way down to the subatomic level. Technically, a right or left bias in nature is called *chirality*. This word, like chiromancy (palmistry), derives from the Greek *cheir*, "hand." Chirality is rampant. Even biological molecules have a natural handedness. It is the asymmetric chirality of protein molecules that creates the helical turns of polypeptide chains in amino acids. The double helix of DNA itself has a bias. It invariably twists to the right, apparently because a precise alignment between strands from the male and female is guaranteed only by a shared handedness bias.

Physicists have even detected right- and left-hand variations in weak nuclear reactions inside the atom. The same set of atoms can also link in various ways, forming spatially different molecules. Although usually only one form is functional, most biological molecules exist in enantiomorphic versions. To visualize how such a mirror reversal might affect the behavior of chemicals, hold your hands in front of you with the palms facing. They are mirror images of each other. If you try to match them up other than facing, they will not align, just as a left or right glove can be worn only on the hand for which it was designed. The same bias occurs in atoms that form enantiomorphic links, creating paired structures that precisely mirror each other. Their designation as L or D, for *levrorotatory* (left-turning) or *dextrorotatory* (right-turning), depends upon the direction in which they deflect a beam of polarized light. This distinction is the origin of the cryptic initial you see on dietary supplements such as L-lysine.

Because of their identical atoms, such compounds behave similarly in most chemical reactions. And yet the metabolism of animals responds to them differently. Consider some beautiful examples of the innate handedness of the world. L-carvone is what we smell in caraway, D-carvone the scent of spearmint. From its right-handed bias, D-sucrose, the sugar from cane, is frequently called dextrose; we and our fellow animals can metabolize dextrose but not its enantiomorphic mirror molecule, L-sucrose. The problem with the sedative thalidomide, which was made available in 1959 and banned only after it caused more than three thousand deformed births, was that the body responds in completely different ways to the drug's left- and right-handed forms. With greater understanding of these issues, thalidomide is now an established treatment for leprosy. "We are in effect machines for processing handed molecules," remarks the British physicist Frank Close.

As fascinating as these discoveries are, so far none answers the question of why, among human beings, there is a consistent majority of right-handers and an equally consistent minority of left-handers. Like almost everything else in the universe, handedness has turned out to be more complicated than we first imagined, surprisingly difficult to sort out from the physics-level bias in the structure of the universe—just as its cultural legacy is entangled in every part of our societies. "Have we convinced ourselves," asks Frank Close, "that the Creation was perfect on nothing more than wish fulfilment as evidence of imperfection and asymmetry is all around us and even within us?"

Nowadays such terms as *left-wing* and *right-wing* are merely buzzwords hurled by political parties to stereotype opponents. They date from the eighteenth century, especially from the days of the French Revolution. The seating plan of the French Estates-General, like that of some other European ruling bodies at the time, dictated that the burghers (the so-called Third Estate, the other two being the nobility and the clergy) sat to the left of the king and the nobles to his right. When the topic of the royal veto came up, those on the left voted against it and those on the right for it. In time left came to represent reform, radicalism, and even internationalism. Right symbolized the opposite—tradition, conservatism, patriotism. The assumption that these ideas were at opposite ends of a spectrum inspired the concept of political centrism. In reality, of course, the parties were never so black and white, and at least since World War I, the old descriptions have no longer been accurate.

Considering that handedness is not the unnatural disability that Plato and John Jackson imagined it to be, our inventive cultural response to it is typical of the brain's tendency to promote minor natural differences into major symbolic ones. Paul Simon cleverly linked natural and symbolic handedness in the 1965 Simon and Garfunkel song "A Simple Desultory Philippic, or How I Was Robert McNamara'd into Submission." The narrator claims that he was almost branded a Communist because he is left-handed.

PLAYING CHESS WITH DEATH

A striking tribute to the weirdness of hands appears in Ingmar Bergman's 1957 film *The Seventh Seal*. The action takes place during the Middle Ages. The protagonist, Antonius Block, returns from the Crusades to find plague ravaging Europe. Death personally comes for him. To delay his departure, Block proposes a game of chess that continues at intervals throughout the movie. On one occasion he reveals his next chess move to a priest, only to learn that the half-hidden figure was actually Death himself. Smiling, Block holds up his right hand, turns it to examine it, and says in awe, "This is my hand. I can move it, feel the blood pulsing through it. The sun is still high in the sky and I—and I, Antonius Block, am playing chess with Death."

Later, outdoors, his torments of faith appear unimportant as he and others loll in the sun over a meal. Life feels precious. Someone passes him a wooden bowl full of milk. "I'll carry this moment between my hands," murmurs Block, "as if it were a bowl filled with fresh milk." He lifts the bowl in a firm grip, lovingly, as if holding the moment is what our hands were always meant to do.

CHAPTER 9

Madonna del Latte

BREASTS AND "THE BREAST"

There are few areas of science more loudly argumentative than
the study of how the breasts of female *Homo sapiens* came to be
the way that they are. "One gets such wholesale returns of conjecture," as Mark Twain said about science in general, "out of
such a trifling investment of fact." Just-So stories about the
breasts range the spectrum from tentative and documented to
radical and unsubstantiated. In all such discussions, it is important to remember two points. First, a great diversity of mammals
feed their young with milk glands. We did not invent this
method, and our mammary glands are not unique because we call
them breasts rather than teats. And second, real breasts on real
women come in an impressive variety of shapes and sizes. "The
breast" is an intellectual construct.

In her book *Mother Nature: Maternal Instincts and How They
Shape the Human Species*, evolutionary anthropologist Sarah Blaffer Hrdy writes, "A cottage industry has grown up busily generat-
ing hypotheses to explain the special buildup of fatty tissue
around an adolescent girl's mammary glands." Hrdy points out
flaws in both what we might call the camel's-hump theory, which
postulates that larger breasts are a good place to store fat for lac-
tation, and in what we might call the target-marketing theory,

which claims that larger breasts are a billboard on which women can advertise to men that they have accumulated a reproductively healthy amount of fat. Unfortunately for this idea, breast size alone is not an accurate indicator of potential milk production; it depends upon the volume of healthy glandular tissue. Furthermore, the body does not burn fat from the breasts themselves unless the mother is so close to starving that her metabolism is forced to call upon all of its stored resources. Bodily systems hungry for fuel will draw first upon the thighs and hips. Some biologists argue that breasts do indeed advertise overall reproductive fitness, but Hrdy points out that often breasts begin developing even before menarche—long before a girl is able to feed offspring, because she is not even able to conceive offspring.

Researchers offer plenty of other rationales for breasts. In the early 1970s, Elaine Morgan, in *The Descent of Woman*, presented breasts as evidence for an aquatic phase in the evolution of our primate ancestors. She argued that after our predecessors lost most of their body hair, breasts provided a handhold for the baby. After all, she insisted, other primate young cling to the chest hair of their mothers while nursing at the small, flattish teats. Morgan also speculated that breasts might serve as flotation devices for these littoral apes. Other scientists have suggested that nipple sensitivity is not a side effect of nursing, that the breasts evolved to their current prominence specifically to become eye-catching pleasure buttons. In the late 1960s, Desmond Morris notoriously suggested in *The Naked Ape* that the buttocks may have been so important as a sexual signal, attracting attention to the genitals and advertising their rear access during our prebipedal days, that the breasts evolved to mimic them once protohumans stood upright. As Morris notes, "Virtually all the sexual signals and erogenous zones are on the front of the body—the facial expressions, the lips, the beard, the nipples, the areolar signals, the breasts of the female, the pubic hair, the genitals themselves, the major blushing areas, and the major sexual flush areas." Morris even points out that the prominent pigment around the nipple highlights the skin's erection during sexual arousal. One of his arguments on behalf of this idea—that *Homo sapiens* alone couples while facing rather than from behind—fell apart when bonobos were observed enthusiastically copulating in the missionary position.

In *The Mating Mind: How Sexual Choice Shaped the Evolution of Human Nature*, Geoffrey F. Miller points out that experts on breast-feeding claim that "there is no correlation between breast size before pregnancy and milk production ability after birth. . . . So, we have to distinguish between mammary glands, which evolved for milk production, and enlarged human breasts, which must have evolved for something else." Fertility cues need not be so blatant; like other creatures, we can get by with more understated signals. Then Miller suggests that breasts may have evolved "not to distinguish girls from women, but to distinguish young women from older women." Why sport a cue that will soon fade and advertise expired fertility? Some biologists argue that visual signs with such built-in expiration dates make sense because what nature seems to have planned for the body is early and frequent reproduction, and the female body loses the ability to reproduce long before the male's does. Not surprisingly, other scientists dispute this argument. Yes, but what about indicators of overall fat reserves? asks Miller. And the game goes on.

> *If I have offended anybody with any of my language, all I can say is "tough titty."*
>
> DOLLY PARTON

Breasts are no less a hot topic now than in the past. Nursing-formula companies try to build a better artificial nipple; couturiers design packages to accent breasts; bras squeeze them together and lift them upward; pornographers use ice cubes to make nipples stand erect; medical researchers analyze cancer cells; and censors argue over how much of a breast ought to be visible on prime-time television. As biologists and anthropologists squabble over why we have the breasts we have, the rest of us possess them, desire them, and exploit them, but seldom do we ignore them.

The history of words for the breasts is as diverse as the theories of their biological genesis. While responding in these different ways, human beings have coined many names for this part of the body. Around the end of the first millennium C.E., ancestors of our word *breast* began to show up in written English. *Brest* referred literally to the mammary glands of both sexes, and the related *breost* already meant the figurative repository of emotions and thought, as in Congreve's much later usage, "Music has charms to soothe a savage breast. . . ." At times even this innocent-sounding word has been forbidden. Saying "breast" aloud would have shocked Victorians; at the time diners requesting the breast meat of a chicken would ask instead for "white meat." Although its origin is disputed, the word *bosom* shows up during the same era, referring to a single breast.

Until the last decades of the twentieth century, with its familiar news stories of breast cancer and breast implants, the word was still verboten in polite society. Nowadays *breast* is the least offensive English word to describe the human mammary glands. Many other terms have negative connotations. Consider *tits*. To describe an animal such as the bushtit, this monosyllable dates back to the fourteenth century, and within a couple of hundred years it described both a horse and a young woman. There were no negative connotations to the term until the eighteenth century, when *tit* began to denote a woman of questionable morals. By the seventeenth century at the latest, *tit* referred to the female breast. Not until the 1930s was frigid weather "colder than a witch's tit," but by the forties the phrase *tit man* already described a male who focused his erotic attention upon this portion of a woman's anatomy. *Boobs*, many women's favorite slang term for the breasts because it sounds playful, dates to the end of the Roaring Twenties. But it evolved from *bub* and *boobie* and *bubbie*, which experts trace all the way to the late 1600s. Some philologists speculate that the latter term evolved from the Latin *puppa*, meaning "little girl" (related to the origin of the word *pupil* for the center of the eye, as described in Chapter 3). *Boob sling* as a synonym for *bra* shows up in the late 1960s. Variations on "boob" include *boobers* and *boobahs*, and cosmetic surgery gave us *boob job* in the 1980s. By the end of the decade, the 1986 Miss America, Susan Akin, was saying publicly, "I'll admit to a boob job." Around the beginning of World War II, *knockers* appears, a term usually limited to breasts large enough to knock together. Shortly afterward, the bullet-bra'd femmes fatales who adorned crime comics in the 1950s caused the genre to be renamed "headlight comics." By the 1980s *hooters*—also usually reserved for large breasts—was becoming common in the United States, and quickly became the name of a restaurant chain featuring buxom waitresses. *Jugs*, short for "jugs of milk," seems to have come out of Australia circa 1920. In *The Right Stuff*, Tom Wolfe describes "young juicy girls with stand-up jugs." This utilitarian term is reflected in other nicknames such as *milk fountains* and *milk shakes*, *creamers*, and *dairies*, a kind of disdainful metonymy in which the woman is reduced to her capacity to nourish offspring.

Such etymological forays are unpredictable. *Titillate*, for the record, does not mean to excite with breasts, as some humorist claimed; it is from the past participle of the Latin for "tickle." Yet the now complimentary adjective *exuberant* originally meant "overflowing udders"—from the Latin *ex-*, "out," and *uber*, "udder." Apparently the term was not applied to large-breasted women.

Maura Spiegel and Lithe Sebesta end *The Breast Book: An Intimate and Curious History* with an impressive six-page list of slang words and nicknames for the breasts. The terms range the spectrum from playful to uneasy. Many stress size: *boulders, life preservers, sandbags, kettledrums* and *bongos, Hindenburgs* and *zeppelins*, and Monty Python's *huge tracts of land*. There are the potentially violent *bullets, cannonballs*, and *torpedoes*. Apparently breasts remind many people of fruits or vegetables, as in *peaches, mangoes, tomatoes*, and the venerable *melons* (with a subset of variations such as *flesh melons*). Other food-related terms include *angel cakes* and *cupcakes*. Some nicknames personify the breasts as animals—*snuggle pups, heifers, chihuahuas, sweater kittens.* Nipples inspire *big brown eyes* and *pink-nosed puppies.*

Why do we invent so many slang terms for the human mammary glands? In her book *Womanwords,* Jane Mills points out that euphemisms exist to allow us, in the words of Webster, to "avoid the direct naming of an unpleasant, painful, or frightening reality." Then she asks a compelling question:

> What is it about breasts that is so threatening? . . . No slang or euphemistic term exists for male breasts, almost as if they don't exist at all. But then, phallocentric heterosexuality has meant that few men have discovered the erotogenic qualities of their own breasts.

VIA LACTIA

As the food that nature provides to so many domestic and wild creatures whose lives influenced early societies, milk long ago achieved an almost godlike symbolic character. Sacrifices of animal or even human milk were made to a number of deities, including the Nymphs and the Muses; to Cunina, Cuba, and Rumina, the three Roman genii who guarded infants in their cradles; and even to Priapus, Greece's upstanding fertility deity. To this day in many places, cow's milk is seen as an essential and virtuous source of nutrients.

However highly rated various kinds of milk might be, no other version could approach our own lofty rank. Women's breasts are the world's sole manufacturer of the only food on Earth expressly designed for human beings. By comparison all other foods—peanut butter, spinach, lobster, cow's milk—are just handy carbon-based life-forms in the gingerbread house of the world, which we nibble as we forage. Breast milk is what the human factory designs for the early maintenance of its own products. By serving as a physical conduit for the mother's antibodies, milk even inoculates the newborn against some of its micro-scopic enemies. No wonder it—and the act of breast-feeding—have been venerated throughout history. Over the millennia, in various places around the world, worshipful sculptors carved a number of goddesses who were so fertile and nourishing that a mere pair of breasts were too few to express their bounty. They were burdened with *many* breasts. There is a statue of Artemis, carved in first-century Ephesus and now in the Izmir Museum, whose entire abdomen is covered with a disconcerting array of heavy, milk-rich breasts. As late as 1746, Linnaeus was featuring a statue with supernumerary breasts on the frontispiece of his *Faunus Suecica*. A photograph of the adventurous American sculptor Louise Bourgeois shows her wearing a similar array of sculpted breasts.

Every night the sky exhibits a powerful reminder of the cul-tural importance of breast milk and breast-feeding. The myth in-volves Hera, queen of the deities on Mount Olympus. She is one of the long-suffering wives of Zeus, boss of the Greek pantheon and a god who will copulate with any woman, man, or beast that strikes his indiscriminate fancy. One of Zeus's many offspring is Hercules, who like Jesus has a mortal mother. To nourish Her-cules at the fountain of immortality, Zeus brings him to Mount Olympus to suckle at the breast of his wife. Unaware that her own husband is the child's father, knowing only that the infant has been abandoned, Hera nurses him. Unfortunately, Hercules sucks at her breast so energetically that droplets of milk spill. In fact, already endowed with godlike strength, the infant sucks so hard that he causes Hera pain. When she pulls him from her breast, milk spurts across the heavens. Its droplets form stars. This is the origin of what came to be called in Latin the Via Lac-tia, the Milky Way.

Sexual hormones influence the body in many odd ways, and one of them is olfactory sensitivity. Estrogen enhances the ability to detect and respond to odor, and testosterone reduces this ability; therefore, in general, girls and women are measurably more sensitive to odors than are boys and men of the same age. Women are also more sensitive to odors just before the onset of menstruation, because progesterone levels drop and permit the estrogen to take charge. Pregnancy has the same result. This curious gender inequality shows up from birth. Female newborns respond to many odors more dramatically than do males. At times they kick, suck, or cry in response to an odor they like or dislike, and often their breathing rate changes. Tests, as yet inconclusive, suggest that some infants may express their response with rudimentary facial expressions, like adults. This sensitivity often grows even stronger over the first several days. Such observations reveal our inborn ability to distinguish odors, and similar responses even during sleep demonstrate the unconscious nature of such action. The French scientist Hubert Montagner describes olfactory bonding as "one of the instruments that enables the baby to practice its scales, as it were, via the succouring contact of its mother's body."

And what, you may well be asking, does all of this have to do with the breasts? "Among their various olfactory talents," writes American neurobiologist Lise Eliot, "young babies are almost uncannily good at recognizing their mother's breast odor." Babies simply do not respond to any other scent with the same enthusiasm and discernment that they show toward the all-important perfume of their own mother's teats. Human newborns prefer an unwashed breast over one that has been cleansed of the birthing mess. Even if they have never smelled breast milk before, babies who have been out of the womb for as little as two weeks instinctively turn toward the breast of a lactating woman rather than the breast of one who is not nursing. The same children prove equally sensitive to the odors of their mother's underarms and even her neck.

What animals we are; even lab rats demonstrate the same preference. Nor is the connection between mammary glands and scent surprising, because the breasts are modified sweat glands.

Most important odors on the bodies of mammals are produced by particular glands for a specific purpose. A dramatic example of this specificity is the short-lived patch that produces a strong odor on the forehead of buck roe deer during their rutting season. Almost every single known mammal exudes a unique scent around the mammary glands to guide the newborn to its nourishment. This inborn awareness is heightened by contact with the mother's skin. In turn the act of holding and nursing her baby influences the mother's milk production, and the crucial and pleasurable touching emotionally nourishes both mother and child.

Feeding young by exuding milk is the standard mammalian procedure, practiced by creatures as diverse as hamsters and grizzly bears, camels and dogs. Even the helpless young of marsupials crawl toward the flat glands of their mothers. Nothing could be more natural than a woman nursing her child, but the symbol-obsessed human brain has not been content to leave it alone as a natural act. In her elegant 1997 book *A History of the Breast,* Marilyn Yalom coolly understates breasts' tendency to inspire symbolic thinking: "Though certain features of breasts seem to be eternal, such as their ability to produce milk and their vulnerability to disease, the meanings we attach to breasts are subject to considerable variation." The metaphorical role of breasts is apparent even in Yalom's title. She writes not of *breasts,* functioning physical aspects of a particular body, but of *the breast,* a symbolic cultural artifact.

Yalom tells the story of the medieval paintings and sculptures known as *Madonne del Latte.* These works, by a variety of artists, were a rather shocking development in the history of Christian iconography. For more than a millennium, *Maria Lactans,* the nursing Mary, occasionally had been shown in artworks. In this new development, however, the tableau was more realistic than ever before, even though the single exposed breast usually looked artificial and attached rather than organic and grown. Presumably this seeming awkwardness was an artistic convention to express Mary's not-quite-mortal nature. For the first time, the Virgin had been humanized by the presence of a realistic infant rather than deified by a miniature religious icon at her breast. Yalom points out that the popularity of this kind of image in fourteenth-century Italy may be related to several events occurring at the time. Famine was widespread. Not only did mortal nursing mothers suffer doubly, but the very image of divine nourishment may have provided much-needed spiritual sustenance. After all, for a century or two the church had been describing itself as a loving mother nursing her faithful children. Variations on this theme occur later. In the sixteenth century, Filoseti dell'Amatrice painted Mary standing upright, holding the naked Infant in her right arm, and feeding the damned in Purgatory with streams of milk shooting from her bared breasts like rays of light.

A particular subset of such nurturing images has its own odd history. In Roman mythology Pistas personified love and affectionate commitment. Some accounts and artworks portrayed her as a fertile young woman offering her milk-rich breast to an aged parent. In the first century c.e., Pliny the Elder recounted a story he borrowed from the historian Valerius Maximus:

> A plebeian woman of low position who had just given birth to a child, had permission to visit her mother who had been shut up in prison as a punishment, and was always searched in advance by the doorkeeper to prevent her carrying in any food. She was detected giving her mother sustenance from her own breasts.

Pliny claimed that upon the site of this noble act the Romans built a temple honoring Filial Affection. Such stories had long been categorized together as Roman Charity, and during the Renaissance the newer religion co-opted them as manifestations of Christian Charity. Often the daughter was portrayed nursing her father rather than her mother—a gender switch that, as Marilyn Yalom points out, taints the image with incestuous overtones. In Matthias Stomer's painting *Roman Charity*, for example, the familial relationship between the old man and the young woman is as unsettling as the vast difference in their ages.

The best-known recent incarnation of this idea appeared in 1939, as the closing image of John Steinbeck's novel *The Grapes of Wrath*. In the final chapter, the battered Joad family are drenched in a rainstorm and eventually find refuge in a barn. For all the poor condition of the family, the barn harbors an even less fortunate individual—a middle-aged man who is starving. Throughout the book the Joads and other Okies have fought against both nature and the equally uncaring work bosses and landlords; Steinbeck's mythic imagery suggests that always the downtrodden have only each other for material and spiritual sustenance. The novel ends with pregnant young Rose of Sharon holding the starving man's head in her lap and nursing him.

One school of thought, arising long after the fall of ancient mythology and yet employing its imagery, responded quite differently to the natural mammalian occurrence of breast-feeding. Although its doctrinaire metaphors thrive in such areas as linguistic theory, nowadays psychoanalysis has been largely discredited as a therapy. For all its encouragement to further exploration, Freud's pioneer map of sexual territory turns out to have been as distorted by ignorance and fantasy as any first survey of a previously uncharted continent. Nowadays Freud himself is frequently portrayed as a duplicitous megalomaniac in the grip of personal obsessions. During the twentieth century, however, no idea of the breast influenced thinking more than his did.

The two ruling gods in the Freudian mythology are the penis and the breast; he gives these areas of the body more attention than he allots to the vagina and the clitoris. Dominant, of course, is the bad-boy organ that Freud happened to have so much trouble with himself. Freud insisted that boys' lives are tortured by fear of castration and that girls cannot sleep for envying their brother's penis. "Yet," writes Marilyn Yalom, "like a half-buried goddess statue, the breast could claim that it had been there earlier and had never lost its power. Freud always acknowledged the significance of the breast, without conceding an inch of penis."

Yalom hilariously spoofs Freud's essays on female sexuality with a brief meditation on how psychology would look through a breast-envy theory.

Like most mythologies, Freudianism needed both male and female deities. Early on, Freud promoted the breast to an almost equally powerful role in his pantheon. By the time of his late work *An Outline of Psychoanalysis,* Freud was declaring with his usual certainty that the breast and breast-feeding constitute "the prototype of all later love-relations—for both sexes." Apples in a poem, a bird biting a woman's mouth in a dream—anything might be "revealed" as secretly addressing breast-feeding and the lifelong influence of this first sensual pleasure. With nimble rhetorical backflips, Freud even declared that all human beings suffer the painful sense of loss of the breast "whether a child has really sucked at the breast or has been brought up on the bottle and never enjoyed the tenderness of a mother's care." Although supposedly regarding the issue as crucial with or without actual breast-feeding, for decades psychoanalysts waited with fountain pen poised for the analysand's reply to a standard question: "Did your mother breast-feed you?"

Freudian descriptions of a breast-feeding-inspired oral phase in early childhood development show up all over the place. Consider the following speculation from a 1970 paper entitled "Underlying Themes in the Witchcraft of Seventeenth-Century New England," by the American historian John Demos.

> The prominence of oral themes in the historical record suggests that the disturbances that culminated in charges of witchcraft must be traced to the earliest phase of personality development. It would be very convenient to have some shred of information to insert here about breast-feeding practices among early New Englanders. Possibly their methods of weaning were highly traumatic. . . .

☉

For a variety of reasons, over the centuries many women have been unwilling or unable to nurse their own children. In this situation a newborn was turned over to a wet nurse, usually a woman of lower social class. Records of wet-nursing appear from ancient Egypt to medieval France to Victorian England. The parents' motivations varied. Sometimes the mother had died or was ill. The ability to afford a wet nurse was also a status symbol at certain times. Moreover, repeatedly throughout history we see the idea that preserving your breasts from nursing would prolong their attractiveness—especially in periods that celebrated small and firm breasts. "It seems extraordinary that so little remains today of this once flourishing aspect of child care," writes the American psychologist Shari L. Thurer in *The Myths of Motherhood*. "Wet-nursing seems to have vanished from the collective European memory, perhaps because we are so ambivalent about it." Among groups who could afford a wet nurse, the practice did not die out until the advent of reliable bottled milk.

For all its prevalence, wet-nursing has long been controversial. "It can be no motherly heart that would, without [medical] reason let another suckle her children while her own breasts dried up," fumed Stephanus Blankaart, an Amsterdam physician in the late eighteenth century. His contemporary and fellow physician, Johan van Beverwijck, demanded, "What mother would not rather suckle her child herself and so be a whole mother so as not to subject her innocent lamb to all these perils?" Jean-Jacques Rousseau was typically certain when he declared, "A mother ought to nurse her child. . . . The duty of women on this point is not to be doubted." Of course, most such remarks were written by men, who have always told women how they ought to behave. What most people agreed upon, at least in ideal terms if not always in practical application, was the importance of the mother's breast-feeding. Without a germ theory of disease or knowledge of maternal antibodies, physicians such as Blankaart expressed the matter more simply: "the child has already accustomed to nourishment within the mother." Dutch genre paintings of the era promote the formerly sacred image of the nursing Madonna to everyday symbols of family devotion. Philippa Pullar describes in *Consuming Passions* how rules for the hygiene and character of the wet nurse were codified: "Bull's *Rules for Choice of Wet Nurses* recommended that all candidates be perused like cows: the soundness of their teeth and gums, the clarity of their tongues and complexions, the firmness of their breasts and above all the goodness of their characters were each to be carefully examined."

In 1906 Ambrose Bierce satirized wet-nursing when he redefined our taxonomic class in his *Devil's Dictionary:* "**Mammalia,** n. A family of vertebrate animals whose females in a state of nature suckle their young, but when civilized and enlightened put them out to nurse, or use the bottle." Few commentators today would state the matter so defiantly, but some maintain that the prevalence of wet-nursing in the past ought to have broad implications for contemporary models of child development and parenting. Consider the comments of Shari L. Thurer:

> Michelangelo, Juliet, Scarlett O'Hara, and Winston Churchill all had wet nurses. This was not an obscure enterprise; it involved most infants born in certain European cities at various times. Such a practice, if benign (and we have no evidence one way or the other), seems to contradict some of the most cherished beliefs about mother love and attachment held by today's psychologists, child experts, and, indeed, almost everyone.

The problem is determining whether wet-nursing was benign. The ability to continue to function in society does not imply that better nurturing would not have helped most of us. A sad number of people suffer lifelong fear of abandonment and a resistance to emotional and physical intimacy, among countless other problems. Thurer's examples are merely meant to demonstrate the prevalence of wet-nursing, but they hardly buttress her case. Michelangelo was a notoriously difficult and troubled man, Winston Churchill distant and highly competitive. Juliet Capulet and Scarlett O'Hara are fictional, but the behavior of each has a dramatic and symbolic resonance. Juliet kills herself at a young age, and Scarlett is hardly a doting mother.

There remains plenty of evidence for the positive influence, on both mother and child, of physical and emotional bonding through breast-feeding. In the original 1973 edition of the groundbreaking *Our Bodies, Ourselves*, the authors from the Boston Women's Health Book Collective refuted the medical rejection of breast-feeding that had gained momentum since World War II. The act was, they argued, "satisfying, sensual and fulfilling . . . a pleasant and relaxing way for both mother and baby to enjoy feeding . . . ," and concluded that it was "an affirmation of our bodies."

There is an outrageous link between the hormone that guides the development of the breasts and a group of chemicals found in plants. The primary estrogen, estradiol, is produced by the ovaries. It is the architect in charge of designing and building the mammary glands and the reproductive tract, and it also influences mating behavior. Estradiol proves to be chemically similar to flavonoids, pigments in plants such as clover and other legumes. As the British science writer Lyall Watson explains it, during World War II Australian farmers' livelihood was threatened by a precipitous decline in sheep fertility. The farmers suspected a hidden problem with a new clover variety and in time named the trouble "clover disease." It turns out that, in order to fight back against the creatures that were eating them, the plants were producing a chemical that mimicked estradiol. Odd as it seems, this appears to be a fairly common circumstance—but what on earth is its purpose? "There is only one thing that clover and sheep have in common," writes Watson. "They are partners in a plant-herbivore relationship: one eats the other, and the other does what it can to avoid being eaten too often." The flavonoids act "as a contraceptive pill, reducing grazing by reducing the number of grazers."

Just as these plants inhibit reproduction by producing a chemical that tricks the predators' body into thinking it is pregnant, so does the act of breast-feeding serve as a form of (not fully reliable) contraception by suppressing the body's natural return to fertility. For centuries knowledgeable women prolonged nursing as long as possible before submitting yet again to the depleting rigors of another pregnancy and childbirth.

A curious side effect of pregnancy's hormonal suppression of menstruation is the ancient idea, which flourished even after the Renaissance, that milk and blood are different manifestations of the same vital fluid. Laurent Joubert, the sixteenth-century physician and author of the *Popular Errors*, imagined that he had summed up the issue: "And is it not the same blood, which, having been in the womb, is now in the breasts, whitened by the vital spirit through its natural warmth?" The American historian Thomas Laqueur points out in *Making Sex: Body and Gender from the Greeks to Freud* that anatomists as early as Leonardo knew that there were no apparent links in the body between the blood vessels and the milk-producing glands. This lack of a connection was ignored; it did not fit the model. "The route from womb to breast," writes Laqueur, "is clearly less relevant than the poetics of milk and blood."

THE SHE-DEVIL FROM HOOTER CITY

The official breast . . . is Barbie's breast.

NAOMI WOLF

Like many other celebrities, Barbie never gets a moment of privacy. After weathering decades of criticism, she could not be blamed if she retired to her Dream House and answered the world's abuse with a tiny plastic middle finger. Artists borrow her

likeness for every agenda. Skeptical observers mock her conspic-
uous consumption and unlikely proportions. Responses to Barbie
range from childish glee over her wardrobe to Kuwaiti officials'
issuing a *fatwa* against her in 1995 as a morally corrupting Western "she-devil." Her history, the growth of her myth, is largely a
story of responses to her oversize breasts. Because of them she became popular enough in her original incarnation to be noticed
by an American toy maker. Barbie symbolizes, literally embodies,
the human obsession with breasts.

In 1952 German cartoonist Reinhard Beuthien launched a
series of panel cartoons in the trashy periodical *Bild Zeitung*. It
starred a young woman named Lilli who was not renowned for
her virtue. As Lilli flounces from boudoir to limousine, her sole
ambition is to catch a millionaire, and apparently she has only
one bait for the trap—her body. The leggy, two-dimensional
Lilli is a Teutonic blonde, usually ponytailed, her dramatic eyes
accented by long lashes, her legs tapering into tiny, almost atrophied feet. But her most striking feature is an absurdly disproportionate chest. Lilli's Marilyn Monroe antics proved popular
enough with men to earn her, like a wish granted by a lascivious
sorcerer, a third dimension. Three years after she first appeared
on paper, a toy company began marketing a molded plastic doll
of Lilli. Her most distinctive trait was the same as in the drawings—a cartoonishly mammiferous body barely clad in easily removable clothing. Obviously the Lilli doll was not designed as a
toy for girls. Like the cartoon character she embodied, she was a
novelty for men, the sort of adolescent sex toy that might have
been advertised in *Playboy*.

Ruth Handler, cofounder of a young California toy company
called Mattel, took her daughter Barbara to Europe with her in
1956. One day they came across a Lilli doll in a shop window.
Later Handler claimed that she had long been trying to persuade
the designers at Mattel to produce an adult doll for U.S. toy en-
thusiasts; she bought three Lilli dolls and took them back home to
California. It was a good career move. Although (male) German

cartoonists and toy designers eagerly pitched in while Lilli was intended to titillate an adult male audience, American men balked
when her counterpart was nominated to amuse little girls. Even
with Lilli in hand as a prototype, the men in the Mattel design department still resisted Handler's enthusiasm. They claimed that
the technical difficulties in creating a more flexible version were
insurmountable and that there was no market for such an item.
Handler was convinced that both objections masked the reality,
which was that the thought of crafting and marketing a doll with
breasts made the men uncomfortable. Apparently breasts were an
appropriate preoccupation for men, but not for little girls who
would in time grow breasts of their own.

It seems that several men in Mattel's design department
agreed with Freud that anatomy is destiny. How could a doll
so obviously designed for adult men be remade into a toy for
little girls? Ultimately, the metamorphosis of the bawdy Lilli into
her innocent American cousin retained the disproportionate
breasts—and resulted in the ongoing brouhaha over them. There
have been a number of solemn studies of the impact of Barbie on
American culture, especially on little girls' opinions of themselves and their own breasts. Critics also decry Barbie's miniature
bathroom scale, which was permanently set at 110 pounds, as a
nail in the coffin of anorexics. Defenders, such as American
writer Yona Zeldis McDonough, impatiently reply, "Is there such
widespread contempt for the intelligence of children that we really imagine they are stupid enough to be shaped by a doll?"
Nowadays the doll bears a lot of existential weight on her plastic
shoulders. McDonough's 1999 anthology entitled *The Barbie
Chronicles* includes essays and poems with such titles as "Barbie's
Gyn Appointment" and "Barbie, Twelve-Step Toy."

Considering how problematic their proportions have been
throughout her career, it is amusing that Barbie's breasts caused
trouble even in the planning stage. The German doll, for all her
Junoesque proportions, lacked nipples. When doll makers in
Japan began working on the transformation of Lilli into Barbie,
again and again they returned the prototype to the United States

with nipples. The men at Mattel patiently filed them off and sent back the streamlined prototypes. Mattel executives feared that the doll was going to be a hard enough sell without absolutely inviting public censure. To admit that women had not only breasts but also nipples was more than 1950s America could bear.

A variety of rationales for Barbie's shape have been advanced over the years. These explanations sound suspiciously like alibis ex post facto, designed to cover the facts without recourse to Barbie's Eurotrash ancestry. Lilli is more of a madwoman in the attic than a skeleton in the closet. Her existence is not a secret; it is just not mentioned within the family. Still, despite their subtle differences, Barbie began as a near clone of Lilli. And this fact raises an interesting question. If heterosexual German men collected Lilli, why don't heterosexual American men collect Barbie? She, too, is a chesty babe in skimpy outfits. One reason is the tone of the marketing; Barbie TV commercials tend to air during Saturday-morning cartoons rather than during Monday-night football. As a result, the sight of a Barbie on the dash of a man's car, or on his mantel, would elicit not lewd winks but eloquently raised eyebrows. In the 1990s Joe Blitman, a Barbie dealer and authority, explained the situation: "Very few straight men collect. I think a lot more straight men *would* collect Barbies, or fashion dolls or whatever, but the societal taboos are so strong that they just can't bring themselves to do it." Whether they collect her or not, heterosexual American men seem to react to Barbie as heterosexual German men did to Lilli. "Straight men almost universally respond to the doll by looking at the chest," remarked Blitman. "They can't get beyond that. It's just Hooter City."

TOPLESS

Breast-feeding requires some degree of exposure of the breast being suckled, and therefore it leads us into the uproar about toplessness. In much of the world, breast-feeding is forbidden in public. This ludicrous prohibition embodies the outraged response to bare breasts, whether erotic or maternal or neither.

We have long designed clothing with the breasts in mind. Some garments barely cover them at all. Bikini-like attire—the minimum acceptable clothing in a culture with prohibitions against revealing the breasts and genitals—appears in artwork many centuries older than we might imagine. A mosaic from the Emperor Maximilian's villa in Sicily depicts female athletes participating in a local sporting event, wearing clothing that would not provoke comment on a New Jersey beach. In many places around the world and throughout history, even glimpses of *part* of the breast have been considered too exciting to male sensibilities and therefore a danger to society. In the eighteenth century, Samuel Johnson finally had to tell the actor David Garrick that he could no longer visit him at the theater: "I'll come no more behind your scenes, David; for the silk stockings and white bosoms of your actresses excite my amorous propensities." Apparently a number of ecclesiastics found their own propensities alarmingly stirred, because in 1904 Pope Pius X forbade Catholic women to wear low-cut necklines in the presence of weak-fleshed churchmen.

Completely bare breasts have always been popular in art, from Praxiteles' sculpture of Aphrodite to Delacroix's painting *Liberty Leading the People*. They also show up in the sort of neoclassical public art that adorns many civic buildings. In 1986 one of Ronald Reagan's sidekicks in buffoonery, Attorney General Edwin Meese, held a press conference to declaim the results of his much lampooned "Pornography Commission." This 1,976-page report spent more than 300 pages describing—in precisely the kind of graphic language that it denounced—the sexual antics of people in writings, photographs, and movies. Nadine Strossen, at the time president of the American Civil Liberties Union, pointed out that Meese's report was rather like the anti-pornography writings of Catharine MacKinnon and Andrea Dworkin, who quote lengthy passages from materials they want to ban. With a similar blindness to irony, Meese pontificated upon the sad state of American morals while standing before a statue of a female Justice who was proudly showing off a naked breast. When *Esquire* ran a photo of this dignified moment in jurisprudence the caption read "Tits and Ass." In 2002 John Ashcroft matched his predecessor's high moral standards by covering the statue.

Although no bras were actually burned during the picketing of the 1968 Miss America Pageant, the demonstration included protests against many literally and figuratively confining inventions of an ossified and worried patriarchal society. The bra was one of them. Ancient Roman statuary portrays women wearing a *strophium*, a beltlike support or restraint around the breasts. *Soutien-gorge*, the French call this item, but its English incarnation is named after the old-fashioned French word for a bodice, *brassière*. By the mid-1930s the word was shortened to *bra*. The concept dates back to a patent in 1885 for "health braided wire dress forms," but supposedly the real ur-bra was made of two handkerchiefs in 1914, when a young woman discovered that her corset was not only uncomfortable but showed over the neckline of her dress. Cup sizing did not appear until 1940. Like the corset, the bra could diminish, accent, or exaggerate the curve of the breasts. The highly publicized abandonment of the bra during the late 1960s and early 1970s seems to have had more to do with its role as an artifact of patriarchy than with, as some commentators claim, Rudi Gernreich's invention of the topless bathing suit in 1964. This article of clothing was a fashion statement of limited appeal, but it symbolizes an era that was busily casting off many of the restrictions that formerly had been taken for granted. "Whereas the packaging of breasts as sexual objects had been the norm in the forties and fifties," writes Marilyn Yalom in *A History of the Breast*, "the unbound bosom of the late sixties represented a form of lawlessness, a deregulation of breasts, which were now permitted to flop about without constraint, and a harbinger of greater license yet to come." Breasts and their owners have more freedom now than in the 1960s, but studies have confirmed that, by literally taking some of the burden away from the breasts' own support muscles, bras help keep breasts firm longer.

The bare breasts of women were quickly commodified at topless car washes, massage parlors, and dance clubs. Some activists opposing such exploitation practiced a kind of rhetorical judo by using their adversaries' own weapons against them. In 1964, for example, sixty women and men paraded through downtown Santa Cruz, California, with their chests bare, to attract media attention to many social issues connected with exploitation of the breasts and the female body. Yalom quotes a former model turned feminist activist, Ann Simonton, who presented the group's views in a rousing speech: "If women's breasts weren't hidden in shame or seen as obscene and wicked, how could Madison Avenue, pornographers, movies and television continue to profit off their exposure? . . . We are saying 'no' to the assumption that our bodies belong to advertisers, beauty contests, pornographers, topless bars, peep shows . . . ad nauseam."

In its sense of a bare-breasted woman, the word *topless* dates only to 1964, and it seems originally to have been applied to a bathing suit introduced in San Francisco. It caught on quickly. Within three years the *London Observer* was referring to topless waitresses. The term is not applied to men nowadays, but it turns out that as far back as 1937 *Time* used it with a male definition: "With another bathing-suit season at hand, local lawmakers are aiming their ordinances at males on the score of topless suits rather than at underclad females."

In his short novel *Mr. Palomar*, Italo Calvino dramatized the issues stirred up by the public display of breasts. Even though this chapter of Calvino's little comedy is entirely about the voyeuristic male gaze rather than about being the female object of attention, it highlights the complicated thought processes behind forming a responsible attitude to bare breasts. The maniacally observant Mr. Palomar takes a vacation on some unnamed beach. For the first half dozen pages, he looks—with microscopic attention worthy of Leonardo—at the movements of waves. Then he strolls up the beach and passes a woman who is sunbathing with bare breasts. With his usual obsessive cross-examination of himself, he responds to the sight:

> Palomar, discreet by nature, looks away at the horizon of the sea. He knows that in such circumstances, at the approach of a strange man, women often cover themselves hastily, and this does not seem right to him: because it is a nuisance for the woman peacefully sunbathing, and because the passing man feels he is an intruder, and because the taboo against nudity is implicitly confirmed; because half-respected conventions spread insecurity and incoherence of behavior rather than freedom and frankness.

These are laudable reasons to glance uncertainly at the horizon. Yet Palomar is indecisive, and he reconsiders the matter from another point of view. Perhaps, by looking away, he is contributing to the notion that publicly displayed breasts are in some way illicit, creating what Calvino calls a "mental brassière" between Palomar's eyes and the woman's breasts. A new course of action is indicated. As he wanders back down the beach on his return, Mr. Palomar gazes with impartiality at sea and boats and coastline and "the swelling moon of lighter skin with the dark halo of the nipple." Then he realizes that his last action reduced the woman's nakedness to a mere part of the landscape, and he convinces himself that he may be inadvertently perpetuating the routine disrespect of calcified male superiority. Resolutely Mr. Palomar turns on his heel and walks past the woman again. This time his gaze lingers briefly but appreciatively on her torso. Then it occurs to him that *again* he might be misunderstood. "But couldn't this grazing of his eyes finally be taken for an attitude of superiority, an underestimation of what a breast is and means, as if putting it aside, on the margin, or in parentheses?" He must go back again.

The outcome of this philosophical farce is inevitable. Deciding that she is being stalked by a geek if not a psychopath, the woman jumps up, covers her thought-provoking, conflict-inspiring mammary glands, and departs in a huff.

THE PILLOW OF PARADISE VERSUS THE MILLSTONE OF HISTORY

It may be unfair to ridicule the swimsuits and haircuts of previous generations, but sometimes it is invigorating to mock our ancestors' equally outmoded assumptions and declarations, because these ideas mothered our own worldview. Nowhere does history look more ridiculous than in men's pronouncements upon the unchanging ideal of an alluring abstraction called Woman. A typically silly proclamation appears in Mervyn Levy's 1962 book *The Moons of Paradise: Some Reflections on the Appearance of the Female Breast in Art*:

> Women are like mountains; it is their form rather than their age, or their place in history that commands our interest. They exist, out of time, and it is only the interpretations which men in their lust, or hatred, or tenderness, have superimposed upon the plain reality of the female body that constricts the sex within the compass of an essentially masculine stream of historic events.

Levy elaborates upon this last remark, firmly knotting the rhetorical noose around his own neck: "A woman stands apart from the turbulent stream of events; she is involved in history only in so far as men compel her to assume a variety of fictional roles." By the time he reveals that these observations are based upon his study of the female breast in art, he seems to be kicking the chair away and strangling on his own egocentrism. And yet he keeps going. Breasts are

> the pillow of paradise, the cacti of the desert, the point of departure, and the haven of return. . . . As a symbol of all that a woman can be, in reality or imagination, in fact or in dreams, the breast is a major aesthetic and psychological obsession of the artist and the poet. A man's sense of the beauty of a woman's breast can carry him as close to the divine as he might hope to aspire in this world.

For an antidote to the adolescent maunderings of Mr. Levy, we can turn to the words of Germaine Greer. In *The Female Eunuch,* she pointed out the reality of being assigned to wear the object of such obsession.

> A full bosom is actually a millstone around a woman's neck. . . . Her breasts are only to be admired for as long as they show no sign of their function: once darkened, stretched or withered they are objects of revulsion. They are not parts of a person but lures slung around her neck, to be kneaded and twisted like magic putty.

If Levy is genuinely so ignorant of the historical influence of women in supposedly masculine history, he embodies the very assumptions that women have been battling over the centuries. There is no denying the reality of Greer's comments. Because the optimist in us grasps at straws, however, there are signs of hope in the very publication of such books. In the final paragraph of her masterful study of "the breast," Marilyn Yalom responds to history and thinks about the future, now that women are beginning to have a stronger voice worldwide:

The breast that may be saved will not be the breast that belonged to our ancestors, for women will have some say about its meaning and use. Just as we have found ways to go braless or topless, to promote more breast-cancer research, to fight for the right to breast-feed in public, to counter the glamorous images of the mass media with more realistic images, so too shall we find new ways of protecting and validating our breasts.

CHAPTER 10

Adam's Navel

One of her sisterhood lugged me squealing into life. Creation from nothing. What has she in the bag? A misbirth with a trailing navelcord, hushed in ruddy wool. The cords of all link back, strandentwining cable of all flesh.

JAMES JOYCE, *Ulysses*

UNTYING THE KNOT

Many parts of the body have prompted tributes reminiscent of medieval blazons, but the navel inspires a unique hybrid of lyricism and humor. Why does this part of the body strike many people as silly? Consider this joke: A woman is meditating in the lotus asana when her contemplation of her own navel reveals what appears to be a screw inside the slit. She gets up and finds a screwdriver and sits back down. When she inserts the screwdriver and unscrews the screw, at first nothing happens, but when she stands up, her rear end falls off.

If it is ridiculous, why is the navel also attractive? Unwrapping a new lover would not be quite as much fun without the first sight of the navel. On a beach, glistening with sweat and flecked with grains of sand, a navel can compete with the seashore around it as a natural object worthy of admiration. In the late 1990s, the navel became a fashion accessory—on, of course, a socially approved abdomen, which usually means relatively youthful and toned. Some pop stars made this part of the body their very logo, and store mannequins, teenage girls, and finally adult women fell in step behind them.

Nor is the allure of the belly button a new idea. An Italian legend claims that tortellini are modeled after the navel of Venus. Thousands of years ago, one of the narrators of the Song of Songs expressed his appreciation for his lover's navel: "a round goblet which wanteth not liquor." Since the time of Solomon, those who have felt the need to comment upon what Sir Thomas Browne called "that tortuosity or complicated nodosity" are legion. The surgeon and essayist Richard Selzer, in his collection *Mortal Lessons*, includes a fanciful take on "the sad small stump of the navel, pathetic twist, all that is left of the primal separation, knotted lest our animus leak from us with an obscene little noise." With disconcerting candor, essayist Phillip Lopate writes, "My belly button is a modest, embedded slit, not a jaunty swirl like my father's. Still, I like to sniff the odor that comes from jabbing my finger in it...." Satirizing some of the excesses of twentieth-century art, Martin Gardner wrote a story in which there is an exhibition of photographs of navels that resemble famous faces.

The navel demonstrates the influence of nature upon culture, and two strikingly different observations are worth quoting at length. First, in his novel *Even Cowgirls Get the Blues*, Tom Robbins holds forth with signature excess:

> The umbilicus serves, then withdraws, leaving but a single footprint where it stood: the navel, wrinkled and cupped, whorled and domed, blind and winking, bald and tufted, sweaty and powdered, kissed and bitten, waxed and fuzzy, bejeweled and ignored; reflecting as graphically as breasts, seeds or fetishes the omnipotent fertility in which Nature dangles her muddy feet, the navel looks in like a plugged keyhole

on the center of our being, it is true, but O navel, though we salute your motionless maternity and the dreams that have got tangled in your lint, you are only a scar, after all. . . .

The second long tribute is in response to an analogy Freud used in *The Interpretation of Dreams*: "There is at least one spot in every dream at which it is unplumbable—a navel, as it were, that is its point of contact with the unknown." Critical theorist Shoshana Felman's analysis of Freud includes an examination of this remark. Breathlessly she emphasizes the navel's ambiguous resonance in our always symbolic brains:

> Why, indeed, does Freud choose to call "navel" the dream's relationship to the unknown? The navel marks the place where the umbilical cord is first attached and then (during delivery) cut off; it marks, in other words, at once the *disconnection* and the *connection* between a maternal body giving birth and a newborn child. . . . The navel of the dream embodies, thus, the way in which the dream is, all at once, *tied up* with the unknown and *disconnected* from its knowledge, disconnected from the knowledge of its own begetting.
> And yet, the disconnection has itself the form of a *knot*: the "navel" weaves into the theory of dreams the fundamental paradox of a *knotted disconnection that, itself, will never be untied.*

The navel does indeed mark both our physical link to the past—to our mothers, their parents, all the way back to our proto-human ancestors and their own simian forebears—and also to our independent existence from them. For women who give birth, it is also a link to the future, the very embodiment of evolution in action. Like the Gordian knot, this paradox of the umbilical cord can only be cut through. The very word *paradox* means unbelievable. In his masterful tour of our mortal coil inside and out, *The

Body Has a Head, Gustav Eckstein invoked the mystery inherent in the navel's being an insignia that memorializes such a mythologically distant event as our own entrance into this world: "Every human being every so often in his bathtub looks down with thoughtfulness at his navel, certificate that he likewise crossed the Rubicon, though he does not believe it really. . . . We believe in our birth no more than in our death."

After all, Tom Robbins is correct: The navel is only a scar. Birth scars us for life. Because of the depth of so-called innies, it is easy to forget that the navel is merely a blemish. Often it is represented as more. Publicity materials for Mina "Margery" Crandon, whose confidence games as a psychic Houdini went to great lengths to expose in the 1920s, included nude and seminude photographs of her emitting ectoplasm from various intimate orifices—including her navel. When Wonder Woman in the TV series of the 1970s changed from her secret identity as Diana Prince into her crime-fighting costume, special effects showed some kind of cosmic force seeming to radiate outward from the vicinity of her navel.

In reality all that the belly button produces is lint, and this only from innies rather than outies. The abdomen's centrally located depression serves as a gathering place for clothing fibers, dead skin, and dust. Even this humble substance has attracted scientific attention. In 2002 the science humor magazine *The Annals of Improbable Research* awarded an IgNobel, one of its annual awards for seemingly trivial scientific research, to Karl Kruszelnicki, a science educator and physics professor at the University of Sydney. Other 2002 IgNobels honored equally urgent topics, including a new Indian way to estimate the surface area of elephants and a German study of the role of the mathematical Law of Exponential Decay in the subsidence of beer froth. Fully as devoted as his international colleagues, Dr. Kruszelnicki seems to have made himself into the world's authority on navel lint. During a yearlong study that he funded himself, Kruszelnicki surveyed almost five thousand people. Two-thirds of the participants found navel lint when they searched for it. One third found their lint to be of a particular color, with blue the most common. Kruszelnicki confirmed that both the color of skin and the color of clothing can affect the hue of your navel lint. The substance also increases in proportion to hairiness. To test his theory that abdominal hairs serve as a conveyor belt for transporting fibers upward from underwear—pulling fibers into the pit and preventing their departure—Kruszelnicki asked participants to shave their navel area and report changes. Most confirmed that navel lint subsequently vanished but returned as hair grew back. Pierced navels, perhaps because they occur most often on relatively hairless young women and because they are often exposed, collect less lint than does any other category. The study also found that navel lint increases with age and weight. "Your typical generator of belly-button lint or fluff," declared Kruszelnicki, "is a slightly overweight, middle-aged male with a hairy abdomen." Like the earwax research described in Chapter 4, the navel-lint project *may* harvest unforeseen benefits, but scientists are not waiting with bated breath.

Although the navel is not really a hole, its similarity to one is reflected in two Old English words. *Nafu*—referring to the nave of a wheel, the center through which the axle runs and from which spokes radiate—is close kin to *nafela*, "navel." Through a long process, *nafu* became *nafogar* and *a nauger* and *an auger*, our word for a tool that bores holes. Navel oranges are so named not because of the scar at the top, where the stem attaches, but because they sport a remarkably navel-like excrescence that actually bears a diminutive secondary fruit.

"Round the umbilicus we rotate," declaimed the Scottish surgeon James Bridie in a 1939 lecture to the Glasgow Southern Medical Society. "We are both centripetal and centrifugal." He was thinking of the etymology above. Bridie joins others in a common theme with his observation that this scar "is all that remains of the stem that bound us to the parental stalk. It is a reminder that we have been plucked and must sooner or later die."

To a biologist the navel is our badge of membership in the vast club of placental mammals. This category includes most of the creatures we see around ourselves in the course of a day, from in-laws to pets to cows in a field and elephants in a zoo. You can see the kinship in our skeletons and teeth and hair—and, by definition, in the infant's dependence upon an umbilical cord that connects it with the placenta in the womb of the mother. The placenta is a thin layer of cells formed of tissue from both the mother and the embryo and attached to the uterine wall. During prenatal development it carries oxygen and food from the blood of the mother to that of the embryo and generously returns to the mother's body laden with excretory waste and carbon dioxide for disposal. One problem with this arrangement, and the rationale for a responsible prenatal diet, is that the cord also transports many other substances, including psychoactive drugs such as nicotine and caffeine. For the first couple of weeks after conception, little seems to be happening to the embryo; few women even realize at this point that they are pregnant. But already the future placenta's tissue is connecting to the mother's blood supply. By the fourth week, the newly formed heart is pumping blood between the yolk sac and the embryo, via the brand-new umbilical cord. The uterus expels the placenta after birth.

The attributes of the two other groups of mammals—monotremes and marsupials—demonstrate the uniqueness of our placenta. Monotremes predate the evolution of placental birth. Their few surviving examples include the echidna, the platypus, and the spiny anteater. The term *monotreme* means "one-hole" and refers to a single conduit for both reproduction and digestion. How far back can these creatures trace their bloodline? A single fact is sufficient answer: They lay eggs. In contrast, marsupials give birth, but to young that are too little developed to survive on their own; they mature for a while in a handy pocket on the mother's abdomen. This group was once represented throughout much of the world, but their few descendants are limited mostly to Australia. Marsupials lack nipples and secrete their milk onto the skin, where the young simply lap it up.

During pregnancy the umbilical cord is a life-support system running from the mother to the fetus. Like a pipeline between an astronaut and the mother ship, the cord snakes out from the mother's placenta and plugs into what will become the fetus's navel. For transporting old and new blood, the cord contains one large vein and, twined around it, two smaller arteries. By the time it is fully developed, the umbilical cord is about fifty-five centimeters (twenty-two inches) long and a centimeter (roughly half an inch) wide. No wonder it has to be trimmed after birth. Otherwise a newborn baby would be burdened with it like the Cowardly Lion carrying his tail. Jane Goodall and other scientists report seeing many chimpanzee newborns still attached to the placenta by the umbilical cord for the first few days after birth. Only occasionally do the mothers break the cord themselves. In time it falls off. So does our own, leaving a complicated nodosity to remind us of our links to the rest of the world.

Like our opinions on everything else, our attitudes toward the navel have not been consistent. Victor Mature, the hero of the Cecil B. DeMille film *Samson and Delilah*, was allowed to expose his heroic navel, but Hedy Lamarr, as the temptress Delilah, was forced to hide her own with scarves and jewels. In the original 1939 version of *One Million B.C.*, Carole Landis wore a leather one-piece swimsuit partially to hide her navel. As late as the 1960s, censors of the television comedy series *I Dream of Jeannie* apparently feared that the sight of Barbara Eden's navel would somehow inflame the viewing audience with lust, and for the first couple of seasons she had to keep it under wraps. More recently photographer Dean Brown reversed this trend when he gave a Barbie doll a navel for his portrayal of her as the Venus de Milo.

In 1971 Desmond Morris published the results of his own informal navel-gazing in his book *Intimate Behaviour*, and he wondered about a question that no one else had bothered to ask: What precisely was Hollywood's rationale for covering the navel? Its display is expressly prohibited in the censors' official guidelines during the middle of the twentieth century, "so that," Morris explains, "harem dancing-girls in pre-war films were always obliged to appear with ornamental navel-covers of some kind." One feeble excuse was that a glimpse of the navel might encourage young children to ask embarrassing questions about sexuality—as if they did not already see their own navels a hundred times a day. Morris goes on to detail some curious research. He noticed that in Western visual representations the navel seems to be changing shape. He estimated that contemporary actresses and models were six times likelier to exhibit a vertical navel than the rounder ones of the past.

> A brief survey of two hundred paintings and sculptures showing female nudes, and selected at random from the whole range of art history, revealed a proportion of 92 per cent of round navels to 9 per cent of vertical ones. A similar analysis of pictures of modern photographic models and film actresses shows a striking change: now the proportion of vertical ones has risen to 46 per cent.

Morris considers the changes in body shape and posture but still cannot account for the seeming evolution of navel shape. Apparently other researchers have not flocked to this fertile topic.

Hollywood's timidity about the belly button reflected a long history of ambivalence, especially concerning two figures prominent in Judeo-Christian folklore. Considering the potent symbolism involved, it is not surprising that a number of commentators over the centuries have expressed their opinions on, or at least played with the idea of, the navels of Adam and Eve. In the mid-seventeenth century, Thomas Browne wrote in *Religio Medici* that "the man without a navel yet lives in me." A character in Joyce's *Ulysses* thinks, "Heva, naked Eve. She had no navel." In a

1504 copper engraving by Albrecht Dürer, the navels of the First Couple are obvious, even featured. Yet three centuries later, in *The Angel of the Divine Presence Bringing Eve to Adam*, William Blake portrayed Eve—fully formed and with Godiva locks falling past her buttocks—conspicuously lacking a navel. Adam faces away from the viewer; his abdomen is not visible.

Surely the most entertaining response to this seemingly trivial theological dispute was that of a Victorian science writer named Philip Henry Gosse. Intelligent and passionately religious, he was also a learned author of scientific papers and popular books. In an era of enthusiasm for natural history, he was a sought-after lecturer. Despite all of these accomplishments, in time he became a laughingstock, and today he is remembered mainly for a book addressing, among other things, the appearance of Adam's belly. Its title comes from the Greek word for navel. *Omphalos: An Attempt to Untie the Geological Knot* was published in 1857, two years before Darwin published *On the Origin of Species*. A half century later, Gosse's son Edmund described the world's response to his father's beloved magnum opus: "He offered it, with a glowing gesture, to atheists and Christians alike. . . . But, alas! atheists and Christians alike looked at it and laughed, and threw it away."

In this curious tome, Gosse expressly rejects evolutionary notions of how nature might have developed to its current state of affairs. He begins by arguing that nature moves in a circle, that every process we observe cycles inevitably around to its starting place again. "When once we are in any portion of the course, we find ourselves running in a circular groove, as endless as the course of a blind horse in a mill." He sums up his circular reasoning by saying, "The cow is as inevitable a sequence of the embryo, as the embryo is of the cow." There is no doubt, he argues, that God created everything precisely as described in the Bible. Unfortunately, this point raises a question: How can God create something that moves in a circle without marring the pattern? "Creation can be nothing else than a series of irruptions into circles," he insists.

Supposing the irruption to have been made at what part of the circle we please, and varying this condition indefinitely at will,—we cannot avoid the conclusion that each organism was from the first marked with the records of a previous being. But since creation and previous history are inconsistent with each other; as the very idea of the creation of an organism excludes the idea of pre-existence of that organism, or of any part of it; it follows, that such records are false, so far as they testify to time.

Elsewhere Gosse addresses the implications of this idea:

It may be objected that to assume the world to have been created with fossil skeletons in its crust—skeletons of animals that never really existed—is to charge the creator with forming objects whose sole purpose was to deceive us. The reply is obvious. Were the concentric timber-rings of a created tree formed merely to deceive? Was the navel of the created man intended to deceive him into the persuasion that he had a parent?

Like cranks before him and after, Gosse coins new terms to categorize his claims. Events that occur in the real world of real time he labels *diachronic*, and the illusory appearance of pre-Creation time he dubs *prochronic*. As Stephen Jay Gould remarks in an essay on Gosse, this argument is "often cited as a premier example of reason at its most perfectly and preciously ridiculous." Gould sums up the flaw in Gosse's plan from our later point of view: "If organisms arose by acts of creation *ab nihilo*, then Gosse's argument about prochronic traces must be respected. But if organisms evolved to their current state, *Omphalos* collapses to massive irrelevance."

Jorge Luis Borges emphasizes two virtues of Gosse's idea: its "monstrous elegance" and "its involuntary reduction of a *creatio ex nihilo* to absurdity, its indirect demonstration that the universe is eternal, as the Vedanta, Heraclitus, Spinoza, and the atomists thought." Borges then points out that Bertrand Russell dryly suggested in the 1920s that perhaps the universe was created only a few minutes ago, and that the past that we remember individually and collectively is all an illusion. The sarcasm did not end with Russell. In a mock sermon, the online religious satire site of the fictional Landover Baptist Church beautifully parodies the labyrinthine contradictions of the Adam's-navel controversy. The author explores the theologically opposed positions of Pre-, Mid-, and Post-Umbilicism. The first group believes that Adam, molded of dust in the image of his Sculptor, received his navel during the sculpting. "With this in mind, one must note that if Adam was indeed created in the image of God in Heaven, that [*sic*] our Lord must also be blessed with a navel. One might also argue that God at one time had to have had an umbilical cord." Careful study of the universe's background radiation suggests a possible birthplace for God near the center of what is thought to be the buzzing residue of the Big Bang itself. Mid-Umbilicism argues that the navel is the scar left when God pulled Adam's rib out through his abdomen. *Au contraire*, cries Post-Umbilicism; we were given navels to mark the Fall.

Many people do not find such satire amusing. Yet back in 1944, people were laughing when U.S. Congressman Carl T. Durham and the House Military Affairs Committee he chaired complained that a Public Affairs booklet entitled "The Races of Mankind" would offend fundamentalist soldiers if distributed. What Durham claimed to find objectionable was a cartoonish illustration by Ad Reinhardt, the artist later famous for abstract expressionist and eventually monochromatic minimalist paintings. The cartoon portrayed Adam and Eve—he fig-leaved and she apparently clad in a two-piece bikini-like sarong—with tiny black ink-dot navels. Not everyone accepted Durham's official excuse for opposing this pamphlet. Skeptics pointed out that a far likelier reason for the ultraconservative Durham to dislike it was a table documenting exactly how much higher Northern blacks scored than Southern whites on military-given intelligence tests. Another reason might be that the author of the pamphlet had been accused of being a Communist.

Predictably, in the fundamentalism-haunted United States, this preoccupation with the navels of Adam and Eve *still* comes up. In 1989 a man named Forrest Mims applied for a position as columnist for *Scientific American*. Mims was a Southern Baptist fundamentalist who rejected evolution, the fully documented unifying concept and central tenet of biology. Because of his religion-founded bias, Mims's potential employers had reservations about hiring him to work at one of the world's foremost science magazines. When science journalist (and all-around intellectual polymath) Martin Gardner interviewed him, Mims said flatly that Adam and Eve had no navels because they had no parents. "He was not sure whether Eve had hair," Gardner later wrote,

> and had not realized that hair bears as much evidence of past events as belly buttons. (Hair is dead tissue that grows from its roots, like teeth and fingernails, in small increments. If God could provide Adam and Eve with hair, teeth, and nails that had no past history, he could just as easily give them navels with no past history.) He admitted being puzzled over the question of whether the first trees had rings.

A final note from a most reliable source. On the cover of a 1997 issue of the supermarket tabloid *Weekly World News*, the headline read ADAM AND EVE HAD A BABY GIRL! Presumably, *she* had a navel.

A BRIEF HISTORY OF NAVEL-GAZING

In 1857 an Englishman named Fitz Hugh Ludlow anonymously published a book entitled *The Hashish Eater*. He patterned it on De Quincey's *Confessions of an English Opium Eater* from a few decades before. It includes psychedelic accounts of synesthesia as feverishly overplayed as the film *Reefer Madness*. One commentator accused Ludlow of "hypertrophy of the imagination," and he also suffered from hypertrophy of the vocabulary. In Chapter XVIII Ludlow uses a word that is worth examining further. He describes a man as "rotund in all the plenitude of corporeal well-being—an Omphalopsychite by necessity, since he found it impossible to look down at all without resting his eyes upon that portion of his individuality tangent to the lower border of the waistcoat." Not for Ludlow to simply report that the man was so fat he could not see below his belly.

An Omphalopsychite is a navel-gazer. Ludlow's reason for capitalizing the term goes back into the history of Christianity. In Eastern churches there is a tradition of monastic life called Hesychasm, the foremost mystical movement in the Byzantine Church. The term comes from *hesychia*, a Greek word meaning "divine quietness," supposedly attained by meditating upon God through incessant prayer. The tenets of Hesychasm were rigorously monastic—ascetic simplicity, submission to a master, and the continual prayer. Commitment to these principles would nourish the seeker's intimacy with God and therefore his ability to walk through the world illuminated by the true light of God's glory. This spiritual focus of the entire being, both soul and body, is called intellectual or pure prayer. The repeated invocation of Jesus is why it is sometimes also called "the Jesus prayer."

Hesychasm was a controversial development that launched a passionate theological battle—over, for example, subtle distinctions of Unity and Trinity. These philosophical details became entangled in the ongoing battle between hard-core proponents of the separate Byzantine Church and Latinists seeking reunion with Rome. During the late Middle Ages, a system gradually developed in which believers were advised to attune their prayers to their breathing and focus their eyes on "the middle of the body." This aspect of Hesychasm served its enemies well. In the first half of the fourteenth century, a Greek monk called Barlaam the Calabrian strongly denounced the notion that any rational concept could accurately, or even metaphorically, express the mystical nature of prayer. In a satirical tirade, Barlaam mocked the Hesychasts as *Omphalopsychoi*, people whose souls were in their navels. This slur did not calm the combatants.

In our own time, there is only one prominent meaning of *navel-gazing*: any time-wasting, narcissistic exercise. The *American Heritage Dictionary* defines the term as "excessive introspection, self-absorption, or concentration on a single issue." The variety of usage in a few recent examples is interesting. In 1998 the Australian essayist Peter Goldsworthy published a collection entitled *Navel Gazing: Essays, Half-truths and Mystery Flights*. The next year the online magazine *Salon* ran Ann Beattie's article on our era's glut of confessional memoirs under the title "Navel Gazing Raised to an Art." In 2000, when England's former Tory education secretary, Gillian Shephard, denounced her own party for wasting time rather than concentrating on winning the next election, she added, "And just sometimes there's a little bit too much time or, shall I say, spare energy for navel gazing." One online magazine pundit even refers to himself nowadays as "resident omphalopsychite." Yes, demoted to lowercase. Barlaam the Calabrian would be pleased.

PART THREE

A Leg to Stand On

CHAPTER 11

Privy Members

Let not thy privy members be
layd open to be view'd,
it is most shameful and abhord
detestable and rude.

RICHARD WESTE,
The Booke of Demeanor and
the Allowance and Disallowance of Certaine
Misdemeanors in Companie, c. 1619

THE COMMOTION

The genitalia are the external reproductive and sexual organs—
the parts of the reproductive system that, unlike the uterus and
gonads and seminal vesicles, we can see from outside. Three different views about the genitals demonstrate the variety of approaches to this universal preoccupation. The humorist Dave
Barry expresses their evolutionary purpose in the subtitle of one
of his books: *How to Make a Tiny Person in Only 9 Months, with
Tools You Probably Have Around the Home.* As we all know, this part
of the body does not seem like it is *merely* a part of the body.
"Then there are the genitals," writes Diane Schoemperlen in her
story-and-picture collection *Forms of Devotion*. "This is the area

where most of the commotion is traditionally concentrated." And Walt Whitman concisely defined the reason that the genitals take up so much more space in our minds than they do on our bodies:

> Urge and urge and urge,
> Always the procreant urge of the world.

Unlike asexual creatures such as amoebas, sexual animals cannot reproduce without the tools for the job. Much of this equipment is a variation on the insert-tab-A-into-slot-B method that we ourselves employ, and the diversity of animals wooing and mating out there hints at the impressive variety of genitalia. No area of biology is more eyebrow-raising than the freaky museum of sexuality across the animal kingdom. Many arthropods, especially insects, possess genitalia so elaborate, so outrageously designed to prevent illicit intermingling, that taxonomists employ them as species-specific characteristics for classification. In this book, however, we must limit our prurient curiosity to our own genitals and those of our nearer relatives.

After conception the human embryo is six or eight weeks down the road toward birth before the body's overworked administrative system gets around to assigning it a gender. Despite our inability to imagine a truly genderless state, the primitive body is hardly lost in limbo during this time. With all of its genetic legacy already in place, it is calmly growing fingers and eyes and feet and heart and lungs—establishing all the major attributes of the human form. But it has not yet decided upon a male or female body. This bit of information reminds us just how little variation there is between our two genders. For decades science-fiction magazines attracted teenage boys via pictures of oversexed monsters from outer space slavering over décolletage on magazine covers, but in reality aliens might have a difficult time telling apart the male and female Earthlings.

The idea of men and women as "opposite" sexes, an ordained dichotomy woven into the fabric of nature, proliferates everywhere from homophobia to the feminine mystique to power rhetoric. Therefore it is bracing to be reminded that nature gets the rest of the body well established before attending to such relatively minor distinctions. Even the presence of hormones controlling aspects of sexuality emphasize its graded spectrum. Testosterone and estrogen and their colleagues exist in both male and female bodies; it is their relative proportions that affect development. A little more or less of either can tip the scales away from "normal" gender signals, resulting in women with hairy chins or men with breasts. Other genetic or developmental snafus can result in a variety of hermaphroditic bodies, combining attributes usually found in only one gender.

Recent studies provide further evidence of the complex process of gender determination in the womb. It is not a sudden act but a development influenced by numerous factors. All female mammals possess two large chromosomes, which scientists have named for their similarity X and X. Males have a large and a small chromosome, called X and Y. This apparently unequal distribution of genetic determinants was discovered in 1905 by a Bryn Mawr research assistant named Nettie Stevens. (Few remember her name because breast cancer cut short her career.) The turn toward either ovaries or testes comes about when a single gene on the Y chromosome activates. For convenience, scientists have dubbed this Sex-determining Region of the Y chromosome the SRY, and the news media like to call it "the maleness gene." If this SRY switch does not turn on (and "switch" is the metaphor that geneticists most often employ for the action), the developing body defaults to a female plan. If it does switch on, a male body results. Alison Jolly, the American primatologist and evolutionary biologist, remarks in *Lucy's Legacy: Sex and Intelligence in Human Evolution*, "It is essentially the only switch gene we know that identifies a large alternate caste of people, without apparent disadvantage in survival or reproduction, who have predictably, multiply different bodies and behaviors from those without it."

The female/male bias in the XX/XY configuration means that,

as Jolly phrases it, "For mammals the female body plan is the default setting." Then she addresses the notions of ideal form superimposed upon the crude reality of this mammalian default mode:

> Does this make females the Ur-mother, the Earth Mother? Are females somehow the way we all should be, the ultimate form, Eve evolved complete, with not a rib but a gene tweaked aside to produce a second string of Adams? Or to take the other tack, should we see the female as inert and unformed clay without the touch of that Master Potter, the Y chromosome, to shape her into an ideal male?

Jolly answers her own question by pointing out that this kind of female default plan represents only what has evolved in mammals. The opposite system exists, for example, in birds. Developing along their own diverse routes since parting from the primordial ancestor they share with mammals, birds have evolved a different system. Males possess the similar genes, dubbed Z and Z, and it is the females who have one large and one small, Z and W. Still more alternatives have evolved independently among insects and other creatures. The most visible result for us, among adult human beings, is the presence of male nipples—vestigial teats present in the embryonic default body that never received orders to grow into breasts.

Early on, the fetus possesses a pair of gonads and undeveloped versions of both the female and male reproductive duct systems, called respectively the Müllerian and the Wolffian ducts. If the body becomes female, the Müllerian duct develops into the female reproductive tract; if male, the Wolffian duct takes over. Remnants of the rejected alternative linger in the body as fossil evidence of our androgynous beginning. Until recently parents had to decide upon both a male and a female name and await birth to learn their baby's gender. Nowadays hormonal tests reveal the fetus's gender long before birth, and ultrasound images can even detect the presence or absence of a diminutive penis. Only in the past few decades have we begun to peer inside the womb and learn the similarities between the genders that result in our lifelong sexual commotion.

VAGINA DIALOGUES

In the late 1940s Simone de Beauvoir wrote in *The Second Sex*, "The sex organ of a man is simple and neat as a finger, but the feminine sex organ is mysterious even to the woman herself, concealed, mucous, and humid, as it is; it bleeds each month, it is often sullied with body fluids, it has a secret and perilous life of its own." She further employs moist imagery even while expressing a surprising distaste for it. "Man 'gets stiff,' but woman 'gets wet'; in the very word there are childhood memories of bed-wetting, of guilty and involuntary yielding to the need to urinate. . . . If the body leaks—as an ancient wall or a dead body may leak—it seems to liquefy rather than to eject fluid: a horrid decomposition." She lists a number of terms that she says many women associate with their sexual organs: *suction, humus, pitch, glue, viscous.*

The ambivalence with which even a feminist such as de Beauvoir discussed the female genitals is part of a sad but long-standing tradition. Writing about her childhood, Gloria Steinem remembers that her mother, while educated and "enlightened" (her own mother had been a prominent feminist), refrained from ever mentioning names for the female genitalia—not just slang terms, but any words at all.

> Thus, whether I was learning to talk, to spell, or to take care of my own body, I was told the name of each of its amazing parts—except in one unmentionable area. This left me unprotected against the shaming words and dirty jokes of the school yard and, later, against the popular belief that men, whether as lovers or physicians, knew more about women's bodies than women did.

It was this conspiracy of silence that the American playwright Eve Ensler was battling when she wrote *The Vagina Monologues*.

Since the work's debut in 1996, featuring the author herself, it has been performed thousands of times in more than twenty countries and translated into as many languages. Prominent actresses flock to volunteer; most performances benefit local women's shelters. In the play Ensler describes the actual vagina as a locus of ongoing violence, but also she employs it as a metaphor for the denied, invisible aspects of women's lives. She based the monologues upon interviews with more than two hundred women. Some characters rhapsodize about pubic hair and the miracle of childbirth, and speculate upon what their vaginas would say if they could speak or what they would wear if they could get dressed; others recount sexual abuse and genital mutilation and the mass rape of Bosnian women. Ensler is herself a victim of childhood sexual abuse, and at the end of one performance, she asked fellow victims to stand. Many in the audience rose.

"I say it," explains Ensler about the repetition of the word *vagina* in her play,

> because I believe that what we don't say we don't see, acknowledge, or remember. What we don't say becomes a secret, and secrets often create shame and fear and myths. I say it because we haven't come up with a word that's more inclusive, that really describes the entire area and all its parts. . . . I was worried about what we think about vaginas, and even more worried that we don't think about them. I was worried about my own vagina. It needed a context of other vaginas—a community, a culture of vaginas.

Many people immediately leaped into the fray to prove Ensler's point. She encountered ferocious opposition to naming aloud this body part common to more than half the human race. Often the people responsible for newspaper and radio advertisements, even for theater answering machines and marquees, censored Ensler's title, toning it down to a benign V. *Monologues* or even *Monologues*. Such a now common word as *fuck* is regularly softened to "the f-word," and *cunt* whittled down to "the c-word," but even a dictionary term such as *vagina* suddenly revealed itself as "the v-word." Much of the response was far worse than timid initialese. The national media conglomerate Entercom Communications banned advertisements for the play on its radio stations in, of all cities, New Orleans—actually invoking "community standards" to protect the innocent home of Mardi Gras. A Catholic church newsletter denounced the play as scurrilous, nasty, and "a porno circus masquerading as 'consciousness-raising,'" and declared that "degrading stuff like this has contributed materially to the advanced stage of moral decadence and corruption to which our society has obviously descended." When the play was announced at his college, one young man—a political-science major at Texas A & M and a member of the Young Conservatives of Texas—proclaimed the *Monologues* vulgar, obscene, radical, "feminist filth," and, inevitably, "leftist propaganda."

All of this uproar over saying a word and discussing the reality it names. In the face of so much opposition, Ensler optimistically said, "As more women say the word, saying it becomes less of a big deal; it becomes part of our language, part of our lives."

The vagina became part of our physical and mental lives because of its essential role in sexuality and reproduction—which, at least among human beings, are no longer always the same topic. The vagina is a muscular organ that consists primarily of a tube about four inches long. It is capable of impressive gripping during copulation and equally impressive expansion and expulsion during childbirth. The ecosystem of the vagina is a prime example of symbiosis in *Homo sapiens*—a mutually dependent relationship between ourselves and other creatures. Simone de Beauvoir described the vagina with aquatic imagery because it is indeed a moist environment. Its inner tissues are similar to the mucous membranes in the mouth. The vagina, like the mouth (although physicians consider the vagina much cleaner), requires a healthy balance of bacteria—in this case lactobacilli, the same bacteria found in yogurt. Apparently the quickest way to upset this ecological balance is by douching, which destroys beneficial bacteria and encourages the growth of less friendly organisms. The pH balance of the highly acidic environment is also thrown off somewhat by semen, which, in order to survive here, is coated in the most alkaline substance that the body produces; but a healthy vagina quickly restores its natural balance. The moist environment, like the warm and sweaty genitals of men, also serves as an ideal incubator for sexually transmitted diseases and other types of infection, which merely adds to the parade of troubles common to this area. An irreverent biologist might point out that, because certain venereal creatures are found exclusively on the human body and must have evolved after it did, surely they rather than we are the pinnacle of creation.

As far as our tour of the body is concerned, the most intriguing aspect of the vagina's biology is the angle of its placement inside the body and the location of its entrance. Unlike most of our cousins, humans have a vagina that is at least as accessible from the front as from the rear. For a long time, we thought that face-to-face copulation was unique to human beings. Indeed, the very term *missionary position* grew out of Western missionaries' attempts to enforce frontal intercourse, because variations already practiced by the invaded natives seemed bestial. Unfortunately for this scenario, scientists have for several years been observing bonobos having sex this way as happily as if they had invented it. Perhaps they did. "The downward-tilting human vaginal canal could have evolved via sexual selection," wrote the American anthropologist Helen Fisher, author of *Anatomy of Love* and *The Sex Contract* and other books. If our ancestral females possessed "a tipped vagina and encouraged face-to-face coitus, her partners could see her face, whisper, gaze and pick up nuances of her expressions." Writing before awareness of bonobo sexual positions, she asked of *Australopithecus afarensis*, the species of the famous Lucy:

> Did Lucy copulate face-to-face? . . . All modern women
> have a downward-tilted vagina rather than the
> backward-oriented vulva of all other primates. Be-
> cause of this tipped vulva, face-to-face copulation is
> comfortable. In fact, in this position the man's pelvic
> bone rubs against the clitoris, making intercourse ex-
> tremely stimulating.

Although the human imagination is preoccupied with the
vagina's role in sexual intercourse and birth, in the same neighborhood, below the clitoris, is the opening of the urethra. Because of its location, to urinate without mishap, most women sit
or squat. Some men try to promote their ability to urinate while
standing into a badge of superiority, and at least one woman
aligns herself with such symbolism. "Male urination really is a
kind of accomplishment, an arc of transcendence," says Camille
Paglia in *Sexual Personae*, and she continues this nonsensical argument at length.

> A woman merely waters the ground she stands on. . . .
> A male dog marking every bush on the block is a graf-
> fiti artist, leaving his rude signature with each lift of
> the leg. Women, like female dogs, are earthbound
> squatters.

Paglia's ranking of one method of urination as emblemati-
cally superior to another demonstrates again the brain's eagerness
to assign meaning to every natural function of the body. In his
straight-faced redneck pose, pornographer Larry Flynt, publisher
of *Hustler*, one-upped even Paglia's deliberately contrarian dis-
course. "Women are here to serve men," Flynt once declared.
"Look at them, they got to squat to piss. Hell, that proves it."

The word *vagina* comes from the Latin for the scabbard or sheath
of a sword, and its survival in English in its original form traces

directly to a quip in a work of art. The Roman playwright Plautus, who lived in the third and second centuries B.C.E., has a character in his comedy *Pseudolus* jokingly ask a woman, "Did the soldier's sword fit your sheath?" By the late seventeenth century, the usage was established in English. Countless metaphors portray both the vagina and the uterus as submissive vessels, often in violent imagery for sexual intercourse—such as the insertion of an aggressive weapon into a passive scabbard.

Gloria Steinem tells of listening to a group of girls from the ages of nine to sixteen trying to coin an ideal collective noun for the various parts of the female genitals. Finally they settled upon "power bundle." This informal nomenclature committee's choice of the word *bundle* reminds us that there is more to this area of the body than just the vagina. As she admits, Eve Ensler uses the word *vagina* to cover more than the reproductive tract of female mammals that leads between the uterus and the outside world. The vagina is only part of the so-called private parts.

Most prominent from the outside is the *mons pubis* or *mons veneris*, the pubic mound or mount of Venus (the goddess of love). Adorned with hair, it covers the pubic bone. The term *vulva*, from the Latin *volva* or *vulva* for womb, now refers collectively to the external female genitalia. "Some studies of desirability," writes Joann Ellison Rodgers in *Sex: A Natural History*, "suggest that the plumper or more pronounced the vulva, the more appealing to men because it signals health, and the ability to grasp and hold the penis more closely during intercourse, increasing tensions, friction, stimulation, and pleasure." Not all biologists accept this theory. These larger, protective outer lips are the *labia majora*; the smaller and more sensitive inner lips, the *minora*. Just inside the opening of the vagina, guarded by the labia, is the *clitoris*, a tiny organ similar to the penis in structure and sensitivity but less avid for publicity. *Pudenda*, in use in English from at least the fourteenth century, refers at times to the external genitalia of either sex, but usually it describes the female version. It derives from a Latin word meaning to be ashamed and is related to *pudency,* a dusty term for bashfulness or modesty; *impudent* actually means shameless.

Slang terms for the female genitals derive from all over the map. Lexicographers propose different possible theories for the origin of the word *beaver.* By the mid-nineteenth century, *beaver* was a slang term for a beard, and one theory proposes that someone decided that the triangle of pubic hair resembles a beard. One male commentator (presumably a lonely forest ranger) actually suggested that, when seen from above, the pubic hair of a woman resembles a swimming beaver. Another possible alternative is that the vulva is called a beaver in reference to the nineteenth-century phrase "busy as a beaver." In short, no one knows. However, the more recent pornographer's term *beaver shot* definitely refers to a photograph of the exposed vulva, as opposed to less explicit soft-core images. Another vulgar term, *pussy,* may not be as zoological in its origin as *beaver.* It is true that since the 1500s cats have been called variations of *pus* and *puskatte.* This derivation, however, may not be the origin of the word *pussy* to describe the female genitalia—and, by extension, to describe sexual intercourse with a woman. By the 1660s the vagina was already being described as *pusse,* related to the Old English word for bag, *pusa* or *posa,* and the Old Norse word for a pouch, *puss.* So the slang term *pussy*'s origin may be similar to that of the standard word *vagina:* referring to a space into which something else is inserted. Incidentally, the character Pussy Galore in the James Bond novel and film *Goldfinger,* who scandalized some moviegoers, has since lent her outrageous monicker to a band.

"The worst name anyone can be called," wrote Germaine Greer in *The Female Eunuch,* "is a cunt." Unquestionably, this is an unpopular word with many women and, like *pussy,* a term of abuse. Anyone calling either a woman or a man a cunt is acknowledged to be resorting to the lowest weapon in the verbal arsenal. This four-letter word has plummeted downward through the social scale in the past several hundred years; in the fourteenth century, it was a recognized standard term for the female genitals. It has since inspired a crude and often vicious gang of other terms, from the coarsely descriptive *cunt hair* for a minute measurement and *cunt tickler* for a mustache to the demeaning metonymy of *cold cunt* for an unresponsive woman and *stingy cunt* for an unavailable one. All of these coinages are stigmatized in dictionaries with phrases such as "usu. considered vulgar." Philologist Eric Partridge once said of *cunt* that, "though a language word, neither colloquialism, dialect, cant nor slang, its associations make it perhaps the most notable of all vulgarisms. . . ."

In her wide-ranging book *Womanwords*, Jane Mills explores the secret history of this forbidden word—to use it in print had been actionable for three centuries prior to its modern resurgence—and points out that no modern dictionary had the courage to even define it until 1965. Now and then authors were forced to nod toward the word's existence. One of the more insulting official references appears—as superstitiously elided as the Tetragrammaton—in the 1811 edition of the *Dictionary of the Vulgar Tongue:*

> C**T: the *xovvos* of the Greek and the *cunnus* of the
> Latin dictionaries; a nasty name for a nasty thing. . . .

Jane Mills comments, "But, presumably, not all males found cunts nasty; the same dictionary defines 'Nickumpoop or Nincumpoop' as 'A foolish fellow; also one who never saw his wife's ****.'" Not everyone finds even the word nasty, including some women who refuse to hide it behind ladylike asterisks. Our era provides numerous examples of pejorative designations for others that have been reclaimed by the very people they were intended to diminish—*queer, nigger, bitch.* Inevitably some activists in the women's movement have worn T-shirts proudly announcing what many men have always feared: CUNT POWER! After Eve Ensler made a disparaging reference to the word *cunt* during a performance of *The Vagina Monologues,* a female member of the audience lectured her at length about the word's virtues.

There are countless other slang terms for the female genitals, from *snatch* to *box* to *muff*. Yet even this array pales beside the variety of individual nicknames. Eve Ensler, for example, lists many affectionate, silly pet names—*pooki, mongo, powderbox, fannyboo, mushmellow, poonani, dee dee, split knish, tamale, dignity, pal, nappy dugout*, and even *Gladys Siegelman*. People apply these epithets to their own and to others' genitals. Apparently John Updike is mistaken in his essay "The Disposable Rocket" when he declares that the sense of "the male body being two" is acknowledged in erotica, "where the penis is playfully given its own name, an individuation not even the rarest rapture grants a vagina." Updike's ignorance of fond nicknames for the vagina may not be surprising in the author of *Toward the End of Time*. In this novel the overtly misogynistic narrator explains his preference for oral and anal sex by confessing, "My own sense of it is that, at age sixty-six, I am still working up to the vagina—that Medusa whose sight turned ancient men to stone, that sacred several-lipped gateway to the terrifying procreative darkness. I was not yet, at three score and six, quite mature enough to face its blood-empurpled folds, its musty exudations." (We will return to the unmanning Medusa in the section on the pubic hair.)

At one point in the *Monologues*, Ensler recounts witnessing her grandchild's birth, during which she observed the changes in her daughter-in-law's vagina. She says it changed from "a shy sexual hole" to, among other metaphors, "a sacred vessel." The notion of the vagina as sacred may be a revolutionary one for our time, but, as Updike's remark above indicates, it is not new. The silencing shame that Ensler and Steinem and Greer discuss is not primordial and even today is not universal. Much of mythology celebrates the human genitals. In Hindu myths the penis, *lingam*, and the vulva, *yoni*, unite in spiritual balance. Pre-Christian Irish religion included a figure called the sheilagh-na-gig, whose icons often were portrayed spreading their labia. As late as the eighteenth and nineteenth centuries, many of these characters could still be found as carved stone adornments on many Irish church buildings and also on some English and Welsh churches. Censorious Christian authorities have since eradicated most of them.

At Kilpeck Church in Herefordshire, you can still see a twelfth-century sheilagh, broadly smiling and with the mesmerizing stare and heart-shaped face of an early Picasso portrait. She is parting her labia with her hands.

Typical early representations of this part of the body include the Venus Impudique, the "Shameless" Venus discovered in France in 1864. It is a three-inch piece of mammoth ivory, a female figure that apparently was carved, sometime around 14,000 b.c.e., in its current state: headless, armless, footless, without any specifically modeled features except the vertical slit of the vagina. The same approach—portraying enough of the surrounding female form to identify the opening as unquestionably vulviform—occurs in three life-size figures found in the cave of Angles-sur-l'Anglin in France, which date to roughly the same period. The so-called Woman of Gabillou, also made in France at about the same time, is a line drawing carved into a cave wall. Without head or hands, it portrays the body of a woman lying on her back as a graceful landscape—the breasts a mountainous horizon, her vulva drawn to resemble a small inlet in the incurving shoreline of her thighs.

In 1972 the United States launched the *Pioneer 10* spacecraft from Cape Kennedy. Its job was to explore the asteroid belt, to send back information about the environment of Jupiter, and ultimately to follow a trajectory into more distant space. Aboard the spacecraft was a gold-anodized aluminum plaque that conveyed greetings from Earth to interplanetary explorers who might encounter the ship as its momentum carried it ever farther from its home planet. Astronomers Carl Sagan and Frank Drake directed the plaque's design. Along with diagrammatic representations of identifying features of our location and era, such as pulsar frequencies and characteristics of the hydrogen atom (the most common atom in the universe), the plaque carried outline drawings of a female and a male adult *Homo sapiens*. Basing her design upon the works of Leonardo and the ancient Greeks,

Sagan's wife, Linda Salzman, made outline drawings of the two human figures. She deliberately portrayed them as panracial. The woman had epicanthic folds on her eyes; the man had a short Afro, prominent lips, and a broad nose. These details survived into the final design, except that in outline the hair looks blond and the skin white, and at some point in the process someone changed the man's hair to wavy.

Salzman portrayed the figures nude. Widely publicized in news media, and also widely pirated without authorization, the *Pioneer* plaque attracted all sorts of attention. Much of it was negative. Citizens who presumably possessed genitals of their own nonetheless complained about this debauched enterprise. When the *Los Angeles Times* published a photograph of the plaque, one reader wrote to the editor, "Isn't it bad enough that our own space agency officials have found it necessary to spread this filth even beyond our own solar system?" A *Catholic Review* article suggested that praying hands would have been more appropriate, to advertise Earthlings' commitment to God. In contrast, one person wrote to the *Los Angeles Times* to parody the complaints about "our sending those dirty pictures of naked people out into space." The writer suggested that the reproductive organs ought to have been covered and replaced with a picture of a stork carrying a bundle. "Then if we really want our celestial neighbors to know how far we have progressed intellectually, we should have included pictures of Santa Claus, the Easter Bunny, and the Tooth Fairy."

Fewer people complained that the plaque represented gender unequally. The man has a penis and testicles, but the woman shows no vertical slit representing the vulva. "The decision to omit a very short line in this diagram," wrote Carl Sagan later, "was made partly because conventional representation in Greek statuary omits it. But there was another reason: Our desire to see the message successfully launched on *Pioneer 10*." He admitted that he and his colleagues may have been unnecessarily timid, because at no point did NASA balk at the nude figures; the myth of government censorship, he insisted, is just that—a myth.

Nonetheless, at this moment, unless it has encountered a meteor or been found by aliens, *Pioneer 10* is still flying through interstellar space, disseminating across the cosmos the terrestrial pretense that the vagina does not exist.

LEONARDO'S INTELLIGENT ORGAN

The penis, far from being an impenetrable knight in armor, in fact wears its heart on its sleeve.

                                                        SUSAN BORDO

Many men would not agree with Simone de Beauvoir that the male genitals are inherently less mysterious and troublesome than the female version. The penis is uncontrollable during unconsciousness, plagued with its own ailments, and almost as busy channeling fluid as its female counterpart. It is awkwardly unprotected and can hide neither its enthusiasm nor its lack thereof. Its very externality, however, its visible responsiveness, makes the penis more familiar to men than the inner recesses of the vagina may be to many women. Indeed, nowadays it is more visible to everyone than it has been in centuries. In millennial culture the penis cavorts in porn films and Viagra advertisements, in the impeachment proceedings of U.S. presidents and the revenge of battered wives.

Samuel Johnson would not be pleased with our unwrapped era. In his monumental dictionary of 1755, the human genitals are conspicuously absent. This troublesome part of the body was cross-indexed in the great man's mind with unclean yearnings and divine vengeance, and it was not a proper topic for "harmless drudges," as Johnson himself defined lexicographers. In the dictionary Johnson moves from *peninsulated* ("almost surrounded by water") directly to *penitence* ("sorrow for crimes"), daintily stepping around the very existence of the male generative organ. (As for the vulva, the *VUL* section ends with the word *vulgarity*, refusing to continue and risk exemplifying the concept.)

Yet most male creatures possess a penis of some sort. All

developmentally normal male mammals have one, although this organ varies as much as the rest of the animalian body. We ought to be grateful that ours lacks the outré bells and whistles of some penises out there in the fetishistic world of animals. Some male creatures' reproductive organs are even equipped with scrubbers for fighting off rival sperm—an unsexy anecdotal argument for nature's obsession with reproduction. The human male's penis, in contrast, is simple and straightforward. The head is called the *glans*, from the Latin for "acorn," as is the head of the clitoris. The glans is densely packed with nerve endings, some of them unique to this area of the body. The lip of the glans is the *corona*, and it, too, is outfitted with distinctive nerves. So is the *frenulum*, the wrinkled tag of skin just below the glans. The *urethra* opens in the end of the glans and transports both urine and semen. Because sperm needs a slightly cooler climate, the *testicles* hang outside the body, in the *scrotum*.

Sometimes the very possession of a penis conflates with all aspects of masculine identity. Of course, the organ does not bestow manliness unless it functions as more than the terminus of the urinary tract. This outdoor water-hose seems to impress Camille Paglia, but it provides little archetypal support for the ego. Only an erection does so. The descriptive term for a man who cannot achieve an erection, even temporarily, is *impotent*— powerless. The penis is not described as impotent; the man is.

The symbolic element in this relationship is evident all around us. Imagery of the power-granting erection appears even in remarks ostensibly about other topics. In the late nineteenth century, when Sir Richard F. Burton published his ornate "translation" of the *Thousand and One Nights* (fancy-dressing the tales of Shahrazad as Edward FitzGerald did Omar Khayyám's *Ruba'iyat*), he assured his Victorian readers that he was giving them an "uncastrated copy of the great original." When India tested nuclear weapons underground in May of 1998, Indian newspapers responded with such comments as, "This proves to the world that we are not a eunuch; we have power in the world." Realizing that anything approaching gender equality would weaken his business empire, Bob Guccione, the publisher of *Penthouse*, famously expressed the symbolic castration he feared when he declared, "Feminism has emasculated the American male."

Apparently feminists have not noticed this debilitating side effect of the quest for equality. "What other feature of the human body," asks feminist cultural critic Susan Bordo appreciatively, "is as capable of making the upwelling of desire, the overtaking of the body by desire, so manifest to another?" And she adds a heterosexual female perspective on the erection: "The penis has a unique ability to make erotic feeling visible and apparent to the other person, a transparency of response that can be profoundly sexually moving and empowering to the one who has stirred the response."

Leonardo da Vinci did not agree with Bordo's assessment of the penis. Half a millennium ago, he was kvetching in private about the troublesome genitals. "The act of procreation and the members employed therein are so repulsive," he moaned to his notebook, "that were it not for the beauty of the faces and the adornment of the actors and the pent-up impulse, nature would lose the human species." As far as we know, Leonardo himself never reproduced. He portrayed the genitalia of men numerous times, but, other than in dissection, drew the vulva only twice—and one of those times sketched a frightening orifice worthy of a Freudian spelunker. "Whereas his interest in women seems confined to the face, the hands, the movements of the bust," remarks French scholar Serge Bramly of Leonardo, "when it comes to young men, he pays more attention to thighs, buttocks, and in general everything from the navel downward."

As an observant possessor of one, Leonardo was quite aware of the behavior of this organ, as he indicates in another long tirade that culminates with a half-hearted celebration:

> About the penis: This confers with the human intelligence and sometimes has intelligence of itself, and although the will of the man desires to stimulate it, it

remains obstinate and takes its own course, and moving sometimes of itself without license or thought by the man, whether he be sleeping or waking, it does what it desires. Often the man is asleep and it is awake, and many times the man is awake and it is asleep. Many times the man wishes to practice and it does not wish to; many times it wishes to and the man forbids it. It seems, therefore, that this creature has often a life and intelligence separate from man and it would appear that the man is in the wrong in being ashamed to give it a name or exhibit it, seeking rather constantly to cover and conceal what he ought to adorn and display with ceremony as one who serves.

In a 1990 paper, "Hominid Bipedality and Sexual Selection Theory," Maxine Sheets-Johnstone proposed that one initial virtue of bipedalism might have been the way it facilitated penile display. This kind of behavior is widespread among our primate cousins. Sheets-Johnstone pointed out that most other primates stand upright for more than a moment only when displaying their penises. However, as Geoffrey Miller argues in his 2000 book, *The Mating Mind*, objections to the display theory include the relatively unimpressive human penis itself (other than its larger size), as contrasted with the colorful equipment shown off by some of our cousins, and the similarity of open-legged sitting, a ubiquitous human posture, to the open-legged penile displays of chimpanzees. And even if any part of this theory is true, Miller adds, "Bipedal genital displays to strangers are now considered a criminal offense rather than a legacy of primate courtship."

The ancient paintings in the Lascaux Cave of France include a small figure of a man wearing both the head of a bird and an impressively large erect penis. He is being attacked by a bison. After the painting's discovery during the Nazi occupation in 1940, people who were quick to denounce all cave paintings as fake pointed out that in published photographs the upstanding bird-headed man has barely any penis at all. Obviously, they argued, the work was not only fake but was under constant revision by its perpetrators. In reality the Vichy-appointed magazine editors had erased most of the shockingly oversized member.

Obviously, male preoccupation with the penis is not a recent innovation. Robert M. Sapolsky, a neurologist and primatologist at Stanford, writes in his candid and imaginative *A Primate's Memoir*, "When male baboons who are getting along well run into each other and want to say howdy, they yank on each other's penises." He compares the gesture to dogs "rolling on their backs to let each other sniff at their crotches. Among male primates, this means trust. All the guys did it to the other guys that they were pals with." Have we come very far from the savannah? Male locker-room antics frequently include grabs at the crotch and slaps on the buttocks. In more than one place in Europe there are Paleolithic carvings of gigantic figures sporting impressive erections. Are they tributes to horny gods or concupiscent aliens? Or are they merely elegant scrawls, a boast from some testosterone-flooded Neolithic adolescent? In one of them, amusingly, the artist portrayed the testicles as two loops beside the rhinoform penis, creating an inverted face reminiscent of the World War II graffiti face of Kilroy. It's a fascinating stylized figure.

In fact, the very word *fascinating* refers to the penis. If we go backward in this word's history, we first encounter the Latin *fascinare*, meaning "to bewitch." Its own antecedent, *fascinum*, referred to both a penis and an evil spell. One word had to carry two such weighty meanings because it was associated with Fascinus, a lesser known deity supposedly in charge of sorcery. His symbol? The penis. The phallus often appears in *apotropaism*, the use of ritual and magic as a preemptive strike against evil. The Romans wore apotropaic phallic amulets to deflect the harmful rays of the evil eye. Worshippers of Priapus begged for more reliable erections and more satisfying sexual intercourse. In contrast, by their worship of a *lingam*, the phallic symbol embodying Siva the Creator, Hindus seek not so much hydraulic assistance as an uplifting union with the world's seminal creativity.

One fascinating place where the symbolic penis shows up is in Christian religious art. Countless paintings and drawings portray the naked genitals of the infant Jesus, but many do not stop with a passive representation. From the fourteenth to the late sixteenth century, hundreds of European artists—no less devout than their fellows—portrayed the Madonna fondling her baby's diminutive genitals. You can see this action in everything from a painting by Paolo Veronese to an etching by Hans Baldung to illuminations in various Books of Hours. The art historian Leo Steinberg, in his book *The Sexuality of Christ in Renaissance Art and in Modern Oblivion,* declares that in these artworks the Christ child's genitals "receive such demonstrative emphasis that one must recognize an *ostentatio genitalium* comparable to the canonic *ostentatio vulnerum,* the showing forth of the wounds."

The divine genitals also show up elsewhere in the history of Christianity. To accommodate dramatic fluctuations in size, the penis is covered in skin that is loose and elastic. This *foreskin* is as thin and sensitive as the skin of the eyelids, and so loose-fitting that it can be drawn back to uncover the glans. The foreskin is also called the *prepuce,* and it is usually under this term that it shows up in religious history. The origin of *circumcision,* the surgical removal of the foreskin, is lost in prehistory. The term derives from two Latin words meaning "to cut around." In boys and men it refers to cutting off the foreskin, a relatively simple procedure that—unlike female genital mutilation—has no negative effect on sexual pleasure after the remodeling. Indeed, after circumcision the naked glans exposes even more sensitive area, now shorn of its protective covering.

Although comedian Robin Williams once described ballet dancers as wearing pants so tight that the audience can discern their religious orientation, many people besides Jews practice circumcision. In the ancient Mediterranean, however, the ritual became so associated with the Hebrews that, according to such commentators as Horace and Martial, many men seeking assimilation into Hellenic culture actually underwent painful (pre-anesthetic) surgery to restore the foreskin. Because Jesus was unquestionably Jewish, he must have undergone circumcision. The Bible specifically mentions that this ritual was performed on the traditional eighth day after his birth, and consequently the Catholic Church celebrates January first as the Feast of the Circumcision. Not that this reference prevented some medieval and Renaissance painters from refusing to portray the Infant scarred by Jewish abomination. Saint Thomas Aquinas and other writers considered the Christ child's circumcision—when the incarnated god first suffered human pain—to be the key moment in the redemption of humanity. If Jesus ascended bodily into heaven, reasoned later commentators, he must have left one small piece of his corporeal form back on Earth—his foreskin. (Never mind that he also left shed hair, eyelashes, skin cells, and thirty-three years' worth of fleshly by-products.) The carnal uniqueness of his prepuce inspired a busy traffic in disease-healing, infertility-curing, impotency-banishing holy foreskins.

In the Middle Ages credulity and its exploitation spawned whole lumber yards of True Cross slivers, far more apostolic gravesites than there were apostles, and countless gallons of milk (over a thousand years old but ever fresh) from Mary's own breasts. Both Luther and Calvin skeptically denounced such con games. Yet the Bible itself, in a primitive moment in the Book of Exodus, demonstrates the magical power of the prepuce. When, for no clear reason, God attacks Moses "to kill him," the prophet's wife (Zipporah in the King James Version, Sephora in DeMille's *Ten Commandments*) somehow protects him by circumcising their son with a sharp stone and then touching the child's foreskin to the feet of Moses.

Innumerable churches claimed to possess either the Holy Prepuce or a holy fragment thereof. One of the many places that professed to own it was the court of Charlemagne, where the grotesque artifact was stored in a Purse Reliquary of the Circumcision that inspired a popular fashion accessory. Saint Agnes of Blannbekin claimed that when taking Communion she envisioned herself swallowing the Holy Prepuce. Although Saint Catherine of Siena didn't go that far, she did insist that the ring she wore was a metamorphosed form of the sacred foreskin. David M. Friedman writes in his cultural history of the penis, *A Mind of Its Own*, "This profusion of Holy Prepuces—all of them fetching high prices on the booming relics market—led to the rise of special connoisseurs and the development of certain tests to determine a specimen's authenticity." He adds with eloquent italics, "The most common of these tests was a *taste* test."

Obviously, with the penis as a link, we can quickly go from the divine to the mundane. Our own era may be as grotesque and superstitious as the Middle Ages, but its more secular preoccupations include a Web site devoted to nothing but a list of nicknames for the penis. The invitation NAME YOUR JOHNSON HERE! is followed by hundreds of terms. Naturally the roll call includes established slang such as *dick, putz, boner, tool, dong, pecker, cock, schlong, peter,* and *johnson*. The latter patriarchy includes both the more formal Mr. *Johnson* and what is presumably his homeland, *Johnson County*. (One joke about the abbreviated member of John Bobbitt claims that after his unscheduled surgery he used the alias "Les Johnson.") A preoccupation with the tool kit shows up in *pile driver, piston, screwdriver,* and *gadget*. A depressing number of terms are militaristic, from the merely pretentious *Adam's arsenal* to the scary *bayonet*, from the *bop gun* to the *sword*. The staff of this genital army includes *Little Colonel, Little General,* and even the *Purple-Headed Warrior*. (In his laughable misogynistic rant *The Prisoner of Sex*, Norman Mailer refers to his penis as "the Avenger.") Names that embrace the testicles as well as the penis include the *family jewels* and the parental *Big Jim and the twins*. Some zoophilic owners have named their genitals after other animals' body parts—but not, oddly enough, after other animals' penises; this grotesque museum cabinet includes the *bald-headed mouse, chicken wing, donkey's ears,* and *elephant trunk.* Entire animals with the dubious honor of also being penis names include *cod, lizard, monkey,* and the legendary sea monster *Kraken.*

Apparently few parents teach their children the real names for their genitals, so you hear little boys prattling echoically about their *wee-wee, pee-pee,* or *ding-dong.* The Internet list includes *wigga-wagga* and *wang dang doodle.* Such euphemisms show up in the oddest places. In 1945, during the summer in which he published *Animal Farm,* George Orwell, that usually clear-eyed satirist of realpolitik, watched a new nursemaid bathe his young adopted son and fretfully asked her, "You will let him play with his thingummy, won't you?"

George Orwell uttering the word *thingummy.* The mind reels.

BOSWELL'S JOHNSON

> *I worked with Freud in Vienna. We broke over the concept of penis envy. Freud felt that it should be limited to women.*
>
> "DR." ZELIG, in Woody Allen's film *Zelig*

Although Samuel Johnson refused to define the penis in the *Dictionary,* his biographer certainly wrote about it in his private journal. Young James Boswell might well have faded into history as just another dissolute future laird whoring around London. Instead he wound up playing Dr. Watson to one of the enduring figures of literary history. Although Johnson was a household name by the time of his death, Boswell preserved the curmudgeon's wit, anxieties, and eccentricities for all time in one of the world's great biographies. Boswell's journal dwells upon Johnson but also records, besides drinking bouts and cockfights and daydreams, some of Boswell's many sexual encounters with women of all social classes. On Thursday, 25 November 1762, for example, he wrote, "I had now been some time in town without female sport. I determined to have nothing to do with whores, as my health was of great consequence to me." Actually, Boswell always seemed to feel that the quickest way to conquer temptation was to give in to it, so that it would go away and leave him alone. Over the years, he acquired many cases of gonorrhea, in locales as far apart as Rome and Dublin.

On this particular evening, Boswell began by hunting up old flames but soon found that they had cooled. Finally, against his repeated vow, he sought a professional: "I picked up a girl in the Strand; went into a court with intention to enjoy her in armour. But she had none." He was referring to a prophylactic. At the time this helpful bit of technology might have been made of sheep's gut, secured with a ribbon. Historians of the period solemnly record that this ribbon could be ordered in one's regimental colors, thereby foreseeing today's rainbow of colored— and even flavored—condoms. Nicknames for prophylactic sheaths permitted the French and English, ever at odds, to fling more insults across the Channel. The French called condoms *les capotes anglaises*, "English hats," and the English called them "French letters."

Boswell "toyed" with the lass, he says, but withstood the temptation to indulge and thereby risk yet another bout of the clap. Then the compulsive diarist could not resist jotting down one more detail: "She wondered at my size, and said if I ever took a girl's maidenhead, I would make her squeak. I gave her a shilling." Seldom has such a compliment been repaid so quickly. Nor can we guess at the woman's sincerity; she may have used the line several times during the evening.

The strumpet's compliment was a species of flattery that many men find reassuring. Throughout history and around the world, countless men have worried that their genital allotment might be inadequate to at least one of its assigned tasks. This anxiety has inspired artworks, religious rituals, surgical procedures, crotches padded with every sort of phallic proxy, a tendency to avoid locker rooms, battalions of feeble jokes, the ingestion of noxious potions made from the stolen private parts of *other* creatures, and an impressive array of instruments and concoctions designed to enlarge the diminutive member.

Lately this unease has even fueled the specialty of penis-enlargement surgery. Neither the American Society of Plastic and Reconstructive Surgeons nor the American Urological Association has been willing to endorse any current method of surgical enlargement, but specialists are gradually refining techniques. Maggie Paley, in *The Book of the Penis*, mentions that "such surgery is considered cosmetic unless a man has a certifiable micropenis. If his erect penis is less than $3^2/_3$ inches long and less than $3^1/_2$ inches in circumference when erect, he may have surgery to lengthen his penis, and most likely his health insurance will cover it." Interested readers ought to confirm this rule with their own insurance agencies.

Apparently many men are convinced that a larger penis would guarantee sex and love and therefore make life more satisfying. The preoccupation is everywhere. The *Kama Sutra* taxonomically ranks men by their penis size, describing them as a size-related hierarchy of animals—hare, bull, and horse. It also ranks women by their vagina size—as deer, mares, and elephants—and suggests not unreasonably that sexual satisfaction is more likely among those women and men compatibly endowed. In the late 1990s, promotional trailers for the American remake of *Godzilla* showed the giant lizard noisily demolishing tiny houses, accompanied by the ad line "Size Matters." A millennial issue of a men's health magazine advised its readers to abstain from masturbation for a few days before going to bed with a woman for the first time, so that the erection would attain its highest angle and most impressive girth for the initial unveiling. Even a quick surf through free Internet porn sites reveals that the most urgent need of the human male is not world peace, love, a meaningful life, or even reversal of baldness. *Does It Swing When You Walk?* asks one ad. Photos show huge cartoonish penises, worthy of Priapus, digitally grafted onto images of smaller bodies. Most of the ads—for exercises, pills, creams—are the worst kind of confidence game, but legions of men are desperate enough to try them. One site includes an amusing pair of before-and-after photographs. In the first a man holds a ruler alongside his penis; in the second the same modest organ is shown with the ruler moved to make the glans lie a couple of inches closer to self-esteem.

There are innumerable jokes about penis size. Cocktail-party theories about Napoleon's urge to conquer inevitably return not only to his diminutive height but also to his less-than-imperial penis. This tidbit of military history may or may not be unrelated to the following anecdote. In his novel *Breakfast of Champions*, Kurt Vonnegut lists the penis size of various characters. Later he told Maggie Paley, author of *The Book of the Penis*, that readers at West Point had complained that, at six and a half inches, the penis of an army officer in the novel was unpatriotically small.

Soldiers, emperors, and other men may find it comforting to learn that *Homo sapiens* has the largest penis of any of the higher primates. About 80 percent of the men measured in studies over the last several decades possess a penis measuring between five and seven inches long when erect; the great majority are six inches or just a bit longer—the length of Vonnegut's general's. The limp penis is closer to half that size, but the smaller the penis, the more it is likely to grow during erection; the length of the unexcited organ does not reliably predict the full erection. Masters and Johnson called the erection "the great equalizer." The average human penis, even flaccid, would look quite formidable alongside the one-and-a-half-inch member of the orangutan or gorilla. Yet Jared Diamond once pointed out that human and primate penises, despite their size difference, are equivalent in function. The orangutan can perform its arboreal intercourse in many positions. "As for the possible use of a large penis in sustaining intercourse," added Diamond, "orangutans top us in that regard too (mean duration fifteen minutes versus a mere four minutes for the average American male)."

☉

A curious side effect of the obsession with size shows up in avowedly heterosexual pornography that emphasizes the extraordinary penis involved rather than the attributes of the women. (Many women pronounce the mutant organs of these actors laughable, grotesque, and impossible to accommodate.) Such fantasizing raises a question: If the average pornhound is worried that he is modestly endowed, why is he willing to torture himself by gazing upon the mythologically impressive member being advertised? The human male's preoccupation with other males' penis size has many complex layers of biology and culture. For one thing, the voyeur will not be *observing* this freak of nature; he will *become* him for the extent of the masturbation fantasy. This act of envious imagination explains the popularity, among male pornography fans, of film stars such as John Holmes and the actor called Long Dong Silver. The latter experienced a moment of limelight outside his own profession in 1991. During the U.S. Senate confirmation hearings for Supreme Court Justice nominee Clarence Thomas, his former colleague Anita Hill testified that Thomas's many inappropriate sexual remarks to her included comparing his penis to that of Long Dong Silver.

The unspoken issue in the Thomas confirmation hearings was the white man's preoccupation with the black man's penis. It reverberates throughout the history of European exploration of Africa and the history of slavery and subsequent resistance to the granting of African Americans' civil rights in North America. Usually this obsession focuses on the legendary size of the African penis. "Whenever a race is called inferior," remarked composer Ned Rorem, "the men are said to have big dicks. All homosexuals have big dicks. All black men have big dicks. All Asians have small dicks." He added that in his own considerable experience as an enthusiastic homosexual, he had not encountered this racial disparity.

White anxiety on this issue was not comforted by a revolutionary volume that was published in 1986—Robert Mapplethorpe's *Black Book*. It was the first major licit volume of nudes of black men in American history. Three years later, when a Mapplethorpe retrospective was scheduled in Washington, D.C., Senator Jesse Helms of North Carolina famously got hot and bothered because this decadent "pornography" was funded—in small part—by the National Endowment for the Arts. (The exhibition was canceled.) The uproar still hasn't died down from Mapplethorpe's photograph *Man in Polyester Suit*, in which an African American man is fully dressed except for his uncircumcised penis, which hangs heavily out of the fly of his pants.

While the white view of the black penis has always been about racial inferiority and animal savagery, usually it manifested itself in a preoccupation with the member's impressive size. In *A Mind of Its Own*, Friedman addresses the heritage of assumptions about the animal nature of black men implicit in the unrestrained sexuality of Mapplethorpe's model:

> The fact that the Man in the Polyester Suit was shown without a head—and thus without a brain—only accentuated the work's self-evident "truth": He is black. He has a large black penis. He *is* a large black penis.

In any penis of whatever ethnic background, the length and girth affect neither the ability to urinate nor, except in extreme cases, the ability to reproduce. The former is determined by a functional urethra and bladder; the latter, assuming a sustained erection in the first place and its insertion into a vagina, by ejaculate volume and sperm count. It would seem that, in a preoccupation with penis size, the sole issue for the man would be his capacity to inspire a return engagement with his fellow performer. The symbolic aspect of the larger penis, however, goes far beyond the desire to please the opposite sex.

Some men, of course, want to impress a member of their own sex, not just for primate locker-room dominance play but with straightforward sexual intentions. It is axiomatic among commentators (with little supporting data) that the homosexual relationship to the penis doubles the trouble. "What is worse than one man in a sexual relationship obsessing on penis size?" runs a gay joke. "Two men obsessing." There is a gay porn magazine called *Inches*.

Among homosexual and heterosexual men, the size and shape of the penis vary along with the size and shape of every other body part. "But size doesn't always count," writes Joann Ellison Rodgers in *Sex: A Natural History*,

> and while men make a big deal about the size of their penises, the fact is that few women will ever get to measure or compare them. Nor do women need to care, despite many women's insistence that they do, because studies incontrovertibly confirm that female pleasure is independent of penis length or circumference. In sum, adult male penises, in all their size variations, are *relatively* about the same from a woman's standpoint.

In other words, as one feminist cultural critic observed, "The porn penis has nothing to do with women." In this regard the male fantasy penis differs from the actual penis. Maggie Paley shamelessly generalizes about male ambition when she says, "If men had their way they would have penises as big as scimitars, or baseball bats, or cannons, with which they could really intimidate each other." But still she makes a good point when she adds, "Luckily, natural 'counterselection' has kept their size within bounds. In evolutionary terms, a penis loses its rationale if it can't fit inside a vagina."

THE SENATOR'S BODY HAIR

We simply cannot leave this area of the body without looking at one badge of adulthood shared by both women and men. Let us begin with a brief foray into etymology. In Latin the word *pubes* referred to the groin, to the hair found on this region of the body, and to adulthood in general. In zoology the term *pubescent* refers

not to sexuality but to the tiny hairs that cover the bodies of insects and some other animals; in botany it designates the hairy
surfaces of leaves and stems. The adjective *pubic* refers to any aspect of the region of the body around the *pubis symphysis*, the
ventral (usually lower; in our case, frontal) union of the two
pubic bones that form the front of the pelvic girdle—the lowest
part of the abdomen, where the thighs meet.

Puberty is the period at which we first become capable of
sexual reproduction. What is noteworthy in this context is that,
both etymologically and physiologically, the term actually marks
the onset of pronounced body hair. As noted in various places
throughout this book, the hormones that create our sexual awak-
ening have other side effects. In girls the hips and breasts grow
larger, and menstruation begins; in boys the voice deepens, and
the shoulders widen. While males and females are developing in
these strikingly different ways, the delighted and confused participants share one intimate milestone: the growth of body hair, especially in their underarms and on their legs and in the pubic
region. The first appearance of this new hair occurs at different
times for different adolescents. In response to the hair's variable
arrival schedule, the Komachi Hair Company in Japan sells a
pubic wig called the Night Flower. It consists of recycled human
hair. Around the turn of the new millennium, the owner of the
company reported that Komachi's most successful month for
sales is June, "bridal season"; the wigs also sell briskly in spring
and early fall, "when the students go on class trips and the girls
have to bathe together." He insisted that Komachi performs a
useful service, assisting young people through the awkward early
stages of adolescence.

Like other bodily hair, pubic hair varies not only in arrival
time but in texture, density, length, and color. Gore Vidal records
that his U.S. Army indoctrination in World War II included
"how to tell our exquisite allies, the Chinese, from our brutish
enemy, the Japanese." During Vidal's training an information officer pontificated helpfully that "the principal difference is the
pubic hair. The Japanese is thick and wiry while the Chinese is

straight and silky." Vidal says that he alone raised his hand to ask "what sly strategies we were to use to determine friend from foe."

Over thousands of generations, the fur that covered the bodies of our ancestors gradually thinned to its current sparseness. This reduction of body hair into widely separated islands has had some interesting consequences. Provident nature thoughtfully assigns most mammals their own particular lice, and three kinds of lice that live upon the human body have evolved adaptations to their favorite turf. *Pediculus humanus corporis* is the body louse. It is similar to those inhabiting the surfaces of our primate relatives, but as our ancestors gradually begat less and less hairy offspring, this louse had to adapt to the reduction of habitat. Like all creatures, body lice need protection from the elements. To grip their mobile cafeteria, they need hairs no more than a millimeter apart. Roger M. Knutson, an American biologist and writer, describes the louse's predicament: "As those forests of hair became sparser and then declined to the scattered copses we now bear, the lice were forced into confined quarters where they probably never saw the lice from the next forest over." Nowadays body lice reside in the nooks and crannies of unwashed clothing, but they can survive only a few days without a host to nibble. A closely related subspecies, *Pediculus humanus capitis*, sometimes inhabits the furry slopes of our noble heads. Head lice are one of the plagues of schoolchildren, and some pharmacies in Israel sell an electrically charged comb designed to zap them. To our shy tenants' surprise, they star in Robert Burns's poem "To a Louse, On Seeing One on a Lady's Bonnet at Church" and in Tennyson's contemptuous description of another author as "a louse in the locks of literature."

Farther south on the body we find the pubic louse, *Phthirus pubis*, impressively adapted to life in a jungle of hair. These six-legged creatures are the infamous "crabs." The pubic louse is brawnier than its alpine relative and requires hairs about twice as far apart. They can find these in the brows and eyelashes and in the armpits, but especially in the pubic hair. As Knutson remarks of these creatures, "They seem designed to be able to hold on more securely than other lice. Why they would need to hold on more securely is worth some thought." Not that they *always* hold on. During any vigorous meeting of pubic hair between two human beings, *Phthirus pubis* is likely to jump ship and be fruitful and multiply in a new environment.

In 2001 the British director Stephen Frears released a new film, *Liam,* that explores the sad life, haunted by religion and poverty, of a seven-year-old boy in Depression-era Liverpool. In one scene, the title character confesses to a priest that he has glimpsed his mother naked. Her pubic hair taints the boy's mother as disturbingly different from the women he has seen naked in old paintings.

Liam's crisis is reminiscent of (and probably a deliberate reference to) a historical speculation, by now almost ossified into assumption, concerning the reason that John Ruskin and Effie Gray never consummated their marriage. The nudge-and-wink joke in the life of the eminent art critic is that Ruskin was horrified by his first glimpse of female pubic hair, the existence of which had never occurred to him. Mary Lutyens and other scholars offer this interpretation of the slightly ambiguous evidence, and it seems a reasonable one. Unquestionably, Ruskin was terrified of female sexuality. Literature nurtured his fantasies of mythic womanhood, and painting and sculpture provided him with an idealized notion of the smooth, hairless female body. Both Ruskin and Gray were sexually ignorant virgins, both victims of repression so complete it seems unimaginable from this distance. Years after their sexless wedding night, Ruskin told a friend that when he undressed his new bride, he found her body "not formed to excite passion." In a letter in 1854, Effie candidly wrote to her parents that Ruskin admitted to her the reason he had resisted sexual intercourse: "that he had imagined women were quite different to what he saw I was, and that the reason he did not make me his Wife was because he was disgusted with my person the first evening. . . ." A character in *The Vagina Monologues* may be summing up Ruskin's problem when she says, "You cannot love a vagina unless you love hair."

Phyllis Rose suggests that even the sight of Gray's real-life breasts might have left Ruskin unwilling (or unable) to perform. Yet breasts—albeit idealized—abound in classical and neoclassical art, whereas pubic hair is difficult to find. Jean-Auguste-Dominique Ingres carefully avoided degrading his highly stylized nudes with pubic hair. This natural embellishment ought to be present below the navel of the *Odalisque with Slave*, but immaculate skin is all that is visible down to the drapery barely veiling her genitals. Eugène Delacroix, Ingres's opponent in the battle of draftsmanship versus brushstrokes, did not omit pubic hair when drawing nudes but arranged models (or sometimes just shadows) to hide it in finished paintings. Gustave Courbet, after such outrages as daring to portray a woman's less-than-ideal thighs, went on to commit other sins against the artistic status quo. They culminated in *The Origin of the World*, a lover's-eye view of a woman's spread legs, displaying not only her pubic hair but even, within its tangles, a scandalous glimpse of vulva. This work, which Courbet painted for the Turkish collector Khalil Bey—and which was not exhibited in public for many years—signifies an advance in realism but also yet another woman reduced to her genitals: Her head is covered with a cloth. (In this regard the painting is rather like Magritte's later painting *The Rape*, in which a woman's face has been replaced by her torso—breasts for eyes, navel for nose, pubic hair for a mouth.) In 2001, it is worth noting, the Tate Britain mounted an exhibition entitled "The Victorian Nude," in which pubic hair simply did not appear. Yet, although he was himself Victorian in many of the ways that we might use the term nowadays, Ruskin was an impressively knowledgeable art critic. How had he missed such works as Hans Baldung's 1517 painting *Death and the Maiden*, which portrays unmistakable pubic hair?

Artists have not been the only group preoccupied with pubic

hair. Sigmund Freud speculated about the origins of ancient myths of the Gorgon Medusa. Instead of hair on her head, she had writhing snakes, and one glance into her eyes would turn a victim to stone. Freud, who seldom got through a half hour without worrying about his troublesome genitals, decided that the Medusa myth represents the paralyzing, unmanning horror that men experience in response to their first glimpse of a woman's vulva and pubic hair—because, God forbid, she "lacks" a penis. In Freud's scenario the pubic hair is so visible and dramatic that it becomes the focus of the image, and the hairs metamorphose into snakes that embody the terrors of sexuality. Freud's determination to promote his own particulars into hardwired universals are at best problematic. In this context, however, Phyllis Rose remarks dryly, "It might be noted that in his later years Ruskin frequently mentioned the Medusa and was obsessed by visions of snakes."

After eleven years of mental illness, John Ruskin died in 1900. The repressed critic had lived into an era of greater freedom of expression in the visual arts; he lasted until two years after the death of Aubrey Beardsley. During his brief tubercular life, Beardsley's own Nouveau portrayals of pubic hair on women in his illustrations of Aristophanes and Oscar Wilde scandalized and titillated a generation. By this time Gustav Klimt was already painting his *Beethoven Frieze*, ostensibly honoring the fin de siècle Beethoven cult, for the fourteenth Exhibition of the Viennese Secession in 1902 (with the features of the Golden Knight said to be based upon those of his friend Mahler). Taking as its alleged theme the redemptive power of art and love, the frieze also portrays the sexuality of women as one of the wild forces that men must resist in their quest for salvation. Considering the history of the Vienna Secession's politics, presumably the work was also a metaphor of the individual's battle against society. But just as Ruskin might have feared, and just as Freud might have predicted, among the figures with highly visible pubic hair were three snake-haired Gorgons.

☉

Not everyone finds pubic hair so terrifying. Often it is lustrous and beautiful and well serves its purpose as an eye-catching, aromatic ornament to the genitals. It can even become a token of affection. During her notorious affair with Byron, the novelist Caroline Lamb—who was already married to Viscount Melbourne, later prime minister—sent her lover a lock of her hair. Historians would not snicker over this gift had the hair come from her head.

Nowadays pubic hair is out of the closet. You can see male and female versions in work by artists from Sylvia Sleigh to Delmas Howe, from Alice Neel to Philip Pearlstein. Still, even such earlier works as Klimt's and Beardsley's have not lost their power to shock. In 1970 an Australian court decided in *Herbert* v. *Guthrie* that Beardsley's illustrations for Aristophanes' play *Lysistrata* were obscene, because they show pubic hair on the women. Based upon his reading of the play and of a footnote to a single translation, one Justice Hart complained that Beardsley was not only obscene but historically inaccurate. He insisted that the women of the time used depilatories to erase their pubic hair. This odd contention has yet to be proven.

If the women of ancient Greece modified their natural ornamentation, they would not have been the first human beings to do so—or the last. On the date that we would now call 24 August 79 c.e., the long-dormant volcano Vesuvius erupted and buried the two thousand residents of nearby Pompeii under ash, mud, lava, rock, and dust. For centuries the city was considered lost forever, and eventually it was practically forgotten. Although the area's historical treasures had been turning up for a century and a half before, it was not until 1860 that systematic excavations began. They were planned and supervised by Giuseppe Fiorelli, the Director of Works for Italy. When a workman's pick struck a hollow in the soil, history began to reveal some of its strangest records of death. The hollow was the body cavity of a human being who had died during the eruption. The body had vanished, leaving a precise model of itself in the surrounding pumice and ash. Realizing the magnitude of his find, Fiorelli filled the cavity with plaster and made a mold. Soon he began careful excavations with the body cavities in mind. He wound up with plaster sculptures of hundreds of figures immortalized in the act of fleeing—shielding their children, packing their jewelry, running down the street.

The most surprising ancient detail recorded in these plaster casts concerns, of all things, pubic hair. One man apparently died during the act of trying to remove his clothing, and as a result we have an intimate glimpse of ancient body fashion. "A curious peculiarity still distinctly traceable," wrote Fiorelli, "is that the hair of the pubes is shaved so as to leave it in a semi-circular form, such as may be observed in the statues, and which has, I believe, been generally supposed to be merely a sculptural convention."

Nor is this quaint fashion merely a historical relic. To this day many members of the clan of *Homo sapiens*, the creature that knits sweaters for dogs and decorates graves with artificial flowers, still shave their pubic hair. Male porn stars trim theirs to accentuate the length of the penis. Female porn stars shave their pubic area to more dramatically expose the labia, partially because some men are sexually aroused by the prepubescent implications of hairlessness. Some women get their pubic area waxed to avoid telltale hairs peeking out from a bikini. Women have been known to shave their pubic hair into a heart shape for Valentine's Day. In an episode of the cable-TV comedy *Sex and the City*, a man shaves a woman's pubic hair in the zigzag shape of a stroke of lightning. This amuses her until, naked one day in a sauna, she encounters another woman who has been similarly marked by the same man.

Televised jokes about pubic hair are a recent development. It is seldom in view. And yet, although in most cultures pubic hair is not considered appropriate for *public* attention, it has long been an essential part of erotica and pornography. During its first couple of decades, the men's magazine *Playboy* managed to admire countless nude women without showing their pubic hair. This strategy required artful poses and frequent touch-ups in the darkroom. This hairless policy lasted until August 1971, when the

rival magazine *Penthouse* published its first full-frontal nude.
Vince Tajiri, *Playboy*'s photo editor at the time, later said that he
had been reluctant to move in the same direction as their rival
publication. And yet, he added, "Hefner started counting the
number of pubic hairs in every copy of *Penthouse*." Observers
called the ensuing rivalry for voyeuristic attention the Pubic
Wars. Like its historical antecedent, this commercial squabble resulted in an extended empire.

The centerfold of the January 1972 issue of *Playboy* wore her
pubic hair proudly. Writing in his book *Bunny: The Real Story of
Playboy*, Russell Miller reported,

> Hefner agonized about taking this step and asked for
> two sets of proofs to be prepared: in one, the girl, Mar-
> ilyn Cole, held an arm demurely across her body to
> conceal her pubic hair; in the other, she held her
> hands at her sides. Only at the last minute, when the
> presses could not any longer be delayed, did Hefner
> okay the full-frontal version.

This crisis of modern journalism may not compare with the decision to publish the Pentagon Papers, but it had its consequences.
As American prisoners of war returned from extended captivity
in Vietnam, they were astonished to find that *Playboy* centerfolds
had suddenly grown pubic hair.

Bob Guccione, publisher of *Penthouse*, claimed that *Playboy*'s
resistance to pubic hair was a denial of the natural: "And if it was
unnatural that way, then it was natural to show pubic hair. I
thought that if we were prosecuted my legal defense would be
based on the fact that it was natural and that it would be unnatural to show it otherwise." This remark was disingenuous at best.
The famously sexist publisher of the famously sexist magazine
has never been accused of promoting the natural. In the 1980s
Guccione denounced the photographs of Lee Friedlander, who
portrays less packaged and genuinely more natural-looking
women. Many of the models in Friedlander's photos shocked

Guccione's tender sensibilities by having hair under their arms and on their legs.

By this time pubic hair was also showing up in movies and the theater. Because nudity alone was a triumph over censorship, often a glimpse of pubic hair seemed an artistic advance when it was really more of a commercial gimmick. Sometimes its appearance was a natural aspect of the scene; sometimes it was not. Still, full-frontal nudity began occurring often enough that in the mid-1970s the elderly J. B. Priestley remarked of contemporary films and theater, "Pubic hair is no substitute for wit."

At times, however, the absence of pubic hair can be more disturbing than its presence. Nowadays pornographers seem to disagree with the character mentioned above from *The Vagina Monologues*, who says flatly that "hair is there for a reason—it's the leaf around the flower, the lawn around the house." In 2000 Richard Desmond, the millionaire that the British press tends to describe as a "porn baron," bought the *Express*. The English writer Jeanette Winterson responded by examining the social messages implicit in a number of the magazines that Desmond published—titles such as *Only 18, Nude Readers' Wives,* and *X-treme.* "For me, the most disturbing element in all the drying sameness," she complained,

> is the shaved pubic hair. How many women actually shave their pubic hair? According to the sex mags, all of us. In real life, the only females without pubic hair are little girls. . . . Are our private parts so scary that we either have to turn them into pre-pubescent fantasies or preparation for the operating theatre?

Many people think so. In his novel *God Bless You, Mr. Rosewater,* Kurt Vonnegut portrays a character who is in perfect agreement with the hairless pubes of classical art—Indiana's Senator Lister Ames Rosewater, the father of the title character. He would have applauded Justice Hart's remarks about the artistic sins of Aubrey Beardsley. The senator's proudest work as a public servant is the Rosewater Law, which settles once and for all a muddle of conflicting definitions: "Obscenity is any picture or phonograph record or any written matter calling attention to reproductive organs, bodily discharges, or bodily hair." It is the last phrase of this sentence that resonates most strongly for Rosewater. He interrogates his son's psychoanalyst, and the doctor asks about the legislator's disgust toward body hair. Senator Rosewater shudders as he recounts the conversation.

> I asked him to kindly get off the subject, that my revulsions were shared, so far as I knew, by all decent men. . . . There's your key to pornography. Other people say, "Oh, how can you recognize it, how can you tell it from art and all that?" I've written the key into law! The difference between pornography and art is bodily hair!

CHAPTER 12

Our Steed the Leg

Man still bears in his bodily frame the indelible stamp of his lowly origin.

CHARLES DARWIN

TWO LEGS GOOD AND BAD

Laetoli is a high, windswept region in Tanzania, on the southern edge of the Serengeti Plain. It is due west of the famous Ngorongoro Crater and only twenty miles south of Olduvai Gorge. The dwindling remnants of what conservationists call "charismatic megafauna" can still be seen in the area—elephants, wildebeests, baboons, antelopes, giraffes. Periodically a relatively mild volcano, Oldoinyo Lengai (Masai for "Mountain of God"), emits a curious kind of ash dense with carbonatite. This substance possesses a property that paleoanthropologists love. It falls rather dry, but when it meets water it turns into a natural cement that hardens into a rocklike consistency—preserving a record of whatever it surrounded. Long ago Oldoinyo Lengai's more short-tempered sister volcano, Sadiman, regularly belched forth a great deal of carbonatite ash. It was doing so on an important day somewhere around 3.6 million years ago. The day was important to us because in the course of a few hours several circumstances joined forces to provide us with one of the early pages in our family album.

Mary Leakey, matriarch of the first family of paleoanthropologists, rediscovered this lost page in 1978. Following earlier discoveries by African colleagues, she found the footprints of two creatures—and possibly a third smaller one—who had walked together in the falling ash of Sadiman. The immortalized footprints seem to belong to the long-extinct hominid *Australopithecus afarensis*, the clan of the famous early hominid nicknamed Lucy. They reveal that even so very long ago there were creatures walking upright across the African savannah, sometimes in tandem and sometimes abreast. The clearly delineated arrangement of the digits, the distribution of weight on the heel and arch and toes, unite to reveal a foot that was remarkably like ours, although smaller. "The Laetoli footprints," declared Leakey, "demonstrate once and for all that man's earliest ancestors walked fully upright with a bipedal, free-striding gait."

The wet ash made such a reliable plaster that in places Mary Leakey even found therein splash marks from the raindrops that turned it into a recording medium. It reveals that the australopithecines were not the only creatures scurrying about on that long-ago day. Trailing among their footprints are the tracks of the extinct horse *Hipparion*. The ash preserved the characteristically dragged footprints of giraffes. It immortalized the tracks of now extinct elephants with downcurving tusks, pygmy giraffes, saber-toothed cats. The record of these wandering creatures further emphasizes the uniqueness of Laetoli. The other mammals walked on four legs. *Australopithecus* walked on two.

In George Orwell's bold satire of Soviet communism, after the animals take over Manor Farm they rename it Animal Farm. Orwell's stand-in for Leon Trotsky—who in time will be exiled by Napoleon, the vicious porker representing Stalin—is a pig named Snowball. Facing widespread illiteracy among his fellow revolutionaries, Snowball condenses his proposed Seven Commandments into what he considers the essential tenet of Animalism: "Four legs good, two legs bad."

The bipeds in question tend to disagree. In the eighteenth century, the German physicist and satirist Georg Christoph Lichtenberg summed up the human view: "It may not be natural for man to walk on two legs, but it was a noble invention." Gustav Eckstein uses a similar word to emphasize the importance of this step in human evolution, and he simultaneously crowns our ancestors' triumph in the standard evolution narrative: "When the first prehumans rose to their hind feet, wobbled for millennia, finally stood upright, achieved the erect posture, it must have been a sorry day but a touch of grandeur was on it." Every step we take relives that moment. And every ballerina's pirouette confirms our belief that, although the other creatures are fine in their own way, we are justified in awarding ourselves Best of Show.

Ever since Darwin many people have envisioned the evolution of life on earth, and especially the slow "rise" of primates toward our own lofty status, as a narrative whose triumphant climax is the narrator of the saga. Both Lichtenberg's adjective *noble* and Eckstein's noun *grandeur* are firmly in this tradition. No sooner had the increasing knowledge of natural science dethroned the origins model in the Book of Genesis than we began replacing it with the kind of scientific myth exemplified by Henry Fairfield Osborn's 1920s title *Man Rises to Parnassus*. Only increasing brain size and the development of the opposable thumb rival bipedalism as climactic moments in these rags-to-riches narratives.

Thanks to the muscles around them, we have the ability to lock the knees in place to hold us upright—automatically, without thinking about it. We do not have to depend entirely upon the strength of the quadriceps, the muscle in the front of the thigh. Gorillas can stand almost upright when necessary, but because their knee does not lock into a supporting position, the upright stance presents a constant strain on their quadriceps. Even at a glance, the posture looks unstable and temporary, like Groucho Marx standing at attention. Not so *Homo sapiens*, as you can observe in any long queue at a bank. We shift our weight from one leg to another for variety, but our knees lock into an upright position the way that the feet of some birds lock them onto branches while they sleep.

Such adaptations are fascinating, but how did they come about? There are numerous theories—some competing, some complimentary—about the origins of bipedalism. Paleontologists and other scientists speculate about the influence of many factors. In all likelihood the first steps of our ambitious forebears were inspired, accommodated, and perfected (so far as they have been, although the system is hardly perfect) by a whole spectrum of influences. Many quadrupeds stand upright briefly to assess danger or reach food. "It is a capital mistake," Sherlock Holmes cautions Dr. Watson, "to theorize in advance of the facts." And yet, like criminal investigation, evolutionary biology must work with the clues available and from them try to imagine the location and nature of corroborative evidence.

So what can we learn about how we came to evolve postprandial strolls and cakewalks and three-legged races and salsa dancing? A moment's thought will reveal the magnitude of this question and the many avenues that must be explored to begin to answer it. To begin with, we have access to the behavior of those creatures most like us structurally and behaviorally—chimpanzees, bonobos, and the other higher primates. We also require the hard evidence of fossil ancestry. Such crucial artifacts steadily accumulate, and not even the holy relics of saints are attended more devoutly. Scientists reverently measure, weigh, photograph, electronmicrograph, and dissect. They prepare cross-sections of bone structure, measure cranial capacity, analyze DNA. Bipedal locomotion demands certain kinds of leg muscles attached to bear the stresses of upright walking, and in the best fossil specimens similar attachment scars are still visible. Modern biomechanics can compare these with the muscles of contemporary human beings. As noted in Chapter 7, the larger muscles of the body are in the limbs, the areas that do the moving and lifting and fighting.

After scientists identified the first fossil ancestors of humans, there erupted an ongoing chicken-and-egg sort of debate about which came first: Did our ancestors first evolve increased brain size or opposable thumbs or the ability to walk on two legs? Or did the traits that constitute our treasured uniqueness develop together? The answer to this question, like so many, had to await further evidence. In 1924 Raymond Dart studied fragments of the skeleton called the Taung Child, a small-brained child's fossil from the mines of diamond-rich Kimberley, South Africa. He pronounced it a new species. (It is now considered a member of *Australopithecus afarensis,* the same species as Lucy.) As a consequence of Dart's discovery, paleontologists began to realize that our ancestors stood upright long before they developed an outsize brain. Other evidence accumulated. It was difficult to argue anymore after Mary Leakey found the footprints at Laetoli. Most scientists agree that these prints, nowadays even more famous than the tracks of strolling dinosaurs, are those of a primate that walked upright at least 3.6 million years ago. There now seems to be little doubt that bipedalism was the first major change in primate shape that led down the long and winding road to Mikhail Baryshnikov and Florence Griffith Joyner.

Theories of the origin of this curious adaptation are legion, but a glance at a few of them will give us an idea of just how complex is the reconstruction of the clues in this detective story. When the birds of Animal Farm protest Snowball's opposition to the two-legged, the dictator smooths their ruffled feathers by explaining that, because their wings are organs of propulsion rather than of manipulation, they, too, are considered four-legged. Snowball adds in explanation of his anti-two-legged stance a line already quoted in Chapter 8: "The distinguishing mark of man is the *hand,* the instrument with which he does all his mischief." Here the pig agrees with some ideas about the evolution of his bipedal enemy. One of the earliest assumptions about the evolution of bipedalism was that it freed the hands for the making and employment of tools and the gathering of foodstuffs. Variations on this theme still appear. In 1981 the American anthropologist Owen Lovejoy proposed that the evolution of bipedality may

have helped encourage pair-bonding in the evolving social structure of *Australopithecus afarensis*. In this scenario hands freed
from locomotory demands become tools for foraging and hunting:
"If your mate is walking upright, he's better equipped to carry
food, and more likely to bring some to you." Lovejoy posited that
monogamy preceded and was actually prerequisite to a male role
of provider. This presumed male dominance is a central, and of
course highly controversial, tenet of the behavioral model.

The *Journal of Human Evolution* published in 1991 a paper by
Peter E. Wheeler, an English physiologist and paleoanthropolo-
gist, that addresses another possible contribution to the evolution
of bipedality. Wheeler proposed that an upright stance reduced
the area of the body exposed to heat *gain* and, by raising the body
to the cooling breezes available above the ground, improved heat
loss. When you stand upright, you expose less of your body to the
hot noonday sun; lowering the surface exposed to evaporation
conserves the water in the body. One of the problems with this
theory is that the first genus for which we find evidence of adaptation to a relatively treeless savannah is *Homo*. Earlier groups, including Lucy's clan, *Australopithecus*, had already been bipedal
while happily basking in the shade of more wooded environments. Possibly the best contribution of thermoregulatory pressures may have been a later nudge toward greater height.

More recently, in 1996, Kevin Hunt proposed what he called
the "postural feeding hypothesis." It suggests that, because most
of the time that apes spend bipedal is employed in feeding, over
time selection for this trait may have encouraged greater reliance
upon bipedalism. He points out that 80 percent of the time that
chimps spend bipedal is devoted to feeding, as they use one hand
to gather fruit and the other to steady their posture. Australopithecines have the kind of torso structure that indicates a body
built for arm hanging, from which clues some paleontologists
infer that they evolved their bipedalism from the same sort of behavior chimpanzees exhibit.

In October 2002 a controversy erupted over the position in
the human family album of *Sahelanthropus tchadensis* (nicknamed

Toumai), a fossil newly unearthed by French paleontologist Michel Brunet. Anthropologist Milford Wolpoff and others examined the scarring on Toumai's skull where neck muscles were attached, in order to determine if the neck's connection to the spine indicated a bipedal stance. Wolpoff stated our view of the animal kingdom when he applied to Toumai the bipedal rule of exclusion: "It did not have human posture, therefore it is not human."

Obviously a two-legged stance is part of our self-definition. Often scientists cannot resist waxing poetic about bipedalism, because it is a fascinating development that changed so many things about the human body. Nowadays, however, few would fall for the assumption implicit in many older evolutionary lineups: that all of those hairy usual suspects shuffling their poor posture toward the side of the page are aspiring to become the upright white guy with the briefcase.

Unfortunately, our ancestors' quadrupedal bodies seem to have been poorly prepared for the daily stress of being upstanding citizens. Bipedality caused an array of ills, from back and neck trouble to the unusual amount of pain human mothers experience while giving birth. Backaches support countless practitioners of medicine and pseudomedicine. American neurosurgeon Frank T. Vertosick describes this problem in his book *Why We Hurt: The Natural History of Pain*. He points out why sciatica is epidemic and discectomies common—because we are walking bipedally with a skeleton designed for quadrupeds. The spinal nerves and the spinal discs are in too close proximity.

> In the human design, our body weight is directed parallel to our spinal axis, a radical change. Using a horizontal spinal architecture for vertical walking is like using a screwdriver to drive nails: the nails may get driven, but a lot of screwdrivers will be shattered in the process. The price we pay for our new posture? Lots of shattered discs.

We have other troubles that derive from our upright stance. The slow development of true bipedal locomotion in the human young contributes to the dangers of a lengthy and dependent infancy. Even in adults the legs frequently walk right into trouble. Unlike the eyes, nostrils, ears, lungs, and arms, the two legs do not provide a helpful redundancy, ensuring fallback options in case of damage to one of the pair. One leg by itself is not particularly useful. Even a sprained ankle results in the whining victim's staggering about on crutches. The more our societies developed past the need to hunt and battle other creatures to survive, the more time human beings have spent inventing rigorous sports to work off some of our excess energy, resulting in the lucrative pastime of sports medicine.

Samuel Johnson conjured a sexist but vivid image when he said in 1763, "A woman's preaching is like a dog's walking on his hinder legs. It is not done well; but you are surprised to see it done at all." Forget Mickey Mouse and Gary Larson's *Far Side* cartoons of cows walking upright when no humans are around. Forget the clothed and strolling Brer Rabbit and Mr. Mole, and the fancy dancing of King Louie in Disney's version of *The Jungle Book*. Outside of fantasy we are the only mammals that get about entirely on two legs, without even resorting to our knuckles. As we have seen in different areas of this journey along the body, a four-legged creature's evolution into a genuinely two-legged creature is not as simple as an office promotion. It requires a whole palette of changes, both dramatic and subtle, that affect the structure and function of many parts of the body. Considering the several negative consequences of walking on our hinder legs, we should probably admit that we do not do it well. But no doubt the other animals are surprised to see us doing it at all.

THE SCARECROW LEARNS TO DANCE

Long ago our ancestors noticed that we alone among the animals stroll about on our hind legs, and ever since then we have been paying attention to this curious ability. In the late nineteenth century, for example, a French physiologist named Étienne-Jules Marey created a number of multiple-exposure photographs of human movement. Inspired by the innovative American photographer Eadweard Muybridge, Marey created scientific works of art that map frozen stages of movement. Yet each photograph of a position is to walking as a note on a musical score is to an actual symphony. The concert of movement that is a walking human being requires the cooperation of many players—bones, muscles, nerves, the eyes, the brain, a reliable system of equilibrium.

As the controversies over the origin of bipedalism demonstrate, one of the most complicated issues in evolution is distinguishing between causes and effects in these many interactions. Numerous factors contributed to our inherited ability to walk down the aisle at the supermarket without leaning on the cart that we're pushing. For this reason the aging process seems to ironically reverse not only the life cycle of each individual but even that of our species. At an early age every human being enacts anew the cultural milestone of bipedalism. Our individual development of this ability is encapsulated in the progress of the Scarecrow in *The Wizard of Oz*. As incarnated by the great song-and-dance man Ray Bolger, the Scarecrow at first can barely stay upright on his own two feet; only assisted by Dorothy does he stand, then stagger, then walk. Finally he dances.

Human infants go through the same progression. Most at least *try* to walk by the time of their first birthday. We consider this action a worthy goal and a photo opportunity, but most other creatures would not be impressed. By the time your child is toddling, most mammals have been running (or flying, or swimming) for months and have already begat offspring of their own. What they have not done, however, is stand for long on only two limbs, much less proceed by trusting that each limb in turn will support the entire body and accurately move it forward. It is a complex action that requires an unconscious mastery of balance, stability, and sheer strength, each of which takes a long time to mature and interact in the infant's body.

The urge to move in this way shows up long before the child is mature enough to actually walk. Like the movement disorder described in the upcoming section on the toes, the urge to move the limbs in a rhythmic pattern is primordial. It resides not in the brain but in the spinal cord, in a neural center called a *central pattern generator* or CPG. "Whether they walk, gallop, or swim," writes neurobiologist Lise Eliot, "most animals move using a similar spinal circuit, which commands alternate flexion and extension among the limb muscles on either side of the body." In general, the older the reflex in our evolutionary history, the earlier it matures. By the age of six months after conception, the fetus in the uterus has developed a walking CPG. Newborns, even if premature, will raise each leg in turn if held so that their feet just touch the floor. "Although this marching behavior looks quite sophisticated," cautions Eliot, "it is not in any way voluntary: even babies born without a cerebral cortex can perform it, and studies of experimental animals have proven that the cortex is not necessary for this most basic form of locomotor development." The reflex disappears for a long time because infants gain weight more quickly than they gain strength and muscular control. Besides, true walking must await a more mature cerebral cortex. In time, while holding on to a parent's hand, the child will pull herself upward. Eventually she will stand alone, toddle, walk away, run in the school gym, and leave home.

The science writer Philip Ball expressed an important point about the complicated procedure of walking in the magazine *Nature* in 2001: "Cyclists, stilt-walkers and drunks know the feeling: you're okay moving, but fall over when you stop." Scientists analyzing the phenomenon of walking have contrived ornate explanations involving the interaction of such forces as friction, slip, balance, and momentum. "The conventional view," writes Ball, "is that we keep our balance when walking as a result of careful muscular control, regulated by physiological balance mechanisms and vision." Unquestionably, our upright locomotion demands a sophisticated and almost instantaneous neural response to a variety of information coming in from different parts of the body—from toes to the inner ear to, normally, the eyes. But new research indicates that when you walk down a street, you are not necessarily employing all of your body's networked mechanisms.

Ball was writing about research by two robotics engineers, Tad McGeer and Michael Coleman. McGeer proposed in 1990 that the action of walking might by itself lend stability to otherwise drunk-footed robots. This would mean that robots could be made to move about without engineers' necessarily trying to imitate the human body's convoluted mechanisms. It turned out that McGeer's prototype robots were actually able to stand, at least with their legs far apart, so the problem was not addressed as expected. In 1997, however, a team led by Michael Coleman at Cornell built a robot that genuinely cannot remain upright unless moving, and in 2001 they followed through with the mathematical model of such locomotion. This wobbly evolutionary advance in robotics was constructed from a children's Tinkertoy set. A 1998 paper by Coleman et al. seems in its title to describe a striking characteristic of *Homo sapiens,* or at least of the toddler version: "An uncontrolled walking toy that cannot stand still."

THE GODDESS WITH BEAUTIFUL BUTTOCKS

The legs begin with a part of the body that we tend to consider a separate entity. As many examples in this book demonstrate, sometimes different aspects of how we look are related to each other in outrageous ways. One such odd link is the relationship between the knee, the buttocks, and the ability to stand upright. A curious side effect of bipedal evolution is that we are almost certainly the only primate that evaluates the rear view of another member of its species and proclaims, "Nice ass." Thanks to our two-legged stance, we are the only primate that really *has* an ass, at least by our definition of the word. From the human point of view, our cousins are woefully assless. If you permit yourself a discreet glance at the derrières of chimpanzees or gorillas, you will find that they have very little curve back there. Their flat, unpadded, callused bums are not designed for a thong bikini. Our bottoms, however, are rounded,

curvy, and well padded—with occasional exceptions, as demonstrated by one actress's suggestion that news media critical of her
diminutive body mass could kiss her "bony ass."

The buttock muscles, the gluteals, are more highly developed
in human beings than in, say, gorillas. This difference is a consequence of upright walking, which demanded a revised transport
system, requiring stronger (and therefore larger) calf and thigh
muscles for pulling the leg backward and forward. With every
stride, our sturdy *gluteus maximus* and its colleagues strain like
galley slaves to help pull the body forward and over the leg. The
maximus—there is also a *medius* and a *minimus*—covers most of
the buttock and is the heaviest muscle in the human body. We
use it to rise from a chair or to take a step. It straightens the leg at
the hip to enable us to walk, run, and stand upright. Lacking this
particular specialized muscle power, gorillas can achieve little
better than a sort of bipedal shamble, except for brief scurries
impelled by their own momentum. One of skeptics' many reservations about Roger Patterson's notorious 1967 home movie
alleging to portray a hiking Sasquatch is that the creature possessed the large curved posterior typical of human beings but
missing in other primates. (The faux creature also had hairy
breasts, a part of the body that is hairless in all higher primates
and most other mammals.)

Ours is not the first era to be preoccupied with, and occasionally alarmed by, the aesthetic appeal of the human bottom—
especially the female bottom. The many manifestations of the
Greek goddess of beauty included Aphrodite Kallipygos, whose
name literally means "beautiful buttocks." There is a celebrated
marble sculpture with this name in the Naples Museum. It is a
near contemporary copy of an original dating from the first century or so B.C.E., which stood in the sanctuary of a temple in
Syracuse dedicated to this famously beautiful goddess and her famously beautiful rear. It portrays an almost nude young woman,
pivoting, the work composed in a spiral composition designed to
be viewed with equal aesthetic pleasure from any angle. Polythe-
ism is so much more entertaining than its monolithic rivals.

The large and crucial buttock muscles attract a lot of attention. Many heterosexual men who do not worship Aphrodite Kallipygos nonetheless focus the majority of their erotic interest on this area of the body in women. They are known in American vernacular as "ass men," to distinguish them from men preoccupied with breasts or legs or feet. Apparently there is no specialized term for a man who finds a woman attractive without fetishizing an area of her body.

Sometimes scientists are able to support their fetishes with grant money, and consequently one of them noticed an interesting point about how a woman's bottom moves when she walks. In one particular way, women have adapted differently than men to an upright stance. The alluring wiggle perfected by Marilyn Monroe, which is less visible on today's leaner actresses and models but still apparent among real women walking around in the world, seems to have an evolutionary origin. Bipedalism in women is slightly tilted because the female pelvic girdle must remain large enough to permit the passage of infants with oversize heads. This widespread preoccupation has its cultural variations but seems to emerge from a natural predilection. Usually female buttocks are more curved than are male. Female human beings have proportionately larger buttocks and upper thighs than do the females of other apes, as well as larger breasts, because of the evolution of increased fat deposits in these areas. There have been many theories about this development. In *The Prehistory of Sex*, Timothy Taylor proposed that an upright posture encouraged the development of larger buttocks on a woman as a sexual signal, because in a bipedal primate the formerly displayed vulva is now hidden. The larger buttock muscles of bipedality would be a prominent billboard on which to display a sexual signal. One problem with this theory is that women also comment on men's asses—probably as often as they comment on any other aspect of the male body—and gay men analyze the subtleties of curve, cleft, and jean fit on their fellow males.

Some biologists have proposed that these physical traits are the result of sexual selection. This is the process by which, over countless generations, the choice of appealing traits in a mate slowly modifies the opposite sex's physique. Nature accomplishes this slow sculpture by selecting for particular traits that then may be passed on to the triumphant suitor's offspring. Male attributes such as beards and the distinctive human penis—which, as explained in Chapter 11, is much larger and more flexible than the penis of the other higher primates—may well have been selected by the opposite sex. "The fact that both human sexes evolved distinctive sexual ornaments," remarks English biologist Geoffrey F. Miller, "shows that both female choice and male choice were important in human evolution." Our bodies not only differ greatly from those of our cousins; our species is also sexually dimorphic, which means that there are striking physical differences between the adult sexes. In contrast, the gender differences between our closer relatives are slight. In 2000, in his book *The Mating Mind: How Sexual Choice Shaped the Evolution of Human Nature,* Miller propounded at length how the sexes may have influenced change in each other's bodies. He also argues that our slowly enlarging brain may have developed some of its astonishing creativity as a response to the mutual give-and-take of mate selection between the sexes. As noted in Chapter 6, in the section on lip size and color, the key differences between boys and girls begin to show up only when they need to advertise that they are capable of reproducing. "Traits that attain their full form only during sexual maturity and sexual arousal," says Miller, "probably evolved through sexual choice."

The passing centuries have witnessed many fashions designed to emphasize rather than flatten the female buttocks. In the sixteenth century, women wore bolsters, wire superstructures, and all sorts of cushions and pads to accentuate the curve of the derrière. By the late 1800s, this arrangement had evolved into the bustle, a wire cage that held the dress so far out from the body that it created an outrageously caricatured rear view. As late as the mid-twentieth century, magazine advertisements for Frederick's of Hollywood cautioned women "Don't Be Flat!" and offered removable pads that rounded the buttocks for "That

Youthful, Natural, Feminine Look." By the end of the century, the fashion had turned to slender women with subtler curves.

Some people do not save all their admiration for the derrières of others. Thanks perhaps to Narcissus, most of us think of vanity as involving the face, yet many people are vain about other parts of the body. Sometimes women and men obviously admire their own posterior region and happily exhibit it in tight jeans or clingy skirts. It is this part of her anatomy that a nude woman seems to be appreciating in a mirror in a sixteenth-century engraving of the Italian School usually called *Allegory of Vanity—Death Surprising a Woman*. The skeletal intruder, bearing an hourglass as his calling card, is able to surprise the woman because she is lost in an ecstasy of admiring her own ass.

Most of us, however, spend more time admiring the departing view of others. Consider the double entendre from Alejandro Escovedo's song "Castanets": "I like her better when she walks away. . . ." Because they are considered erotic without being genital, the buttocks have always been popular in the visual arts. Perhaps the best-known tush in painting is the one in the exact center of François Boucher's famous nude portrait of Mademoiselle O'Murphy—one of the many women who gained a measure of power by amusing the profligate and dissolute Louis XV in eighteenth-century France. She lies prone as if captured in mid-wiggle, with the curve of her naked buttocks providing the painting's unmistakable focal point and raison d'être. It seems likely that Boucher was thinking as he mixed his oils, "Now, *that* is a nice ass." It is the kind of deliciously sexy scene that inspired Diderot to condemn the painter for moral laxity.

In a similar manner, a particular character from mythology is almost always represented with her naked rear to the viewer. Inspired by a Greco-Roman relief sculpture, a number of Renaissance painters and sculptors portrayed the same view of the Three Graces. This group, although not always a trio, formerly were the Greek Charites and the Roman Gratiae, personifications of beauty and grace who inspired artistic works. In most representations, such as those by Raphael and Correggio, they rest their arms on each other's shoulders or around each other's waists; the woman on each end faces the viewer, while the woman in the middle faces away, usually with the back of her entire nude body visible. You can see the epitome of this theme in the British Museum, in Antonio Canova's neoclassical marble from the early nineteenth century. Three dimensions, however, freed Canova from custom; his three Graces face forward, revealing their idealized posteriors from the same direction. Greek mythology's callipygian trio still attract attention. A late-1990s poem by Jon Erickson incorporates several fictional book titles, including *The Ass of the Middle Grace*.

There is no question that body fashions and the degree of erotic interest in this region vary around the world. One such variation is *steatopygia*, a medical term from the Greek words for *fat* and *rump*. In the film *One-Trick Pony*, Paul Simon uses the word to mock a man who mistakenly thinks it is flattering. Sometimes whole cultures exhibit steatopygia, notably large and protruding buttocks, and such exaggerated features are the focus of not only erotic interest but even certain aspects of social rank. (A corresponding concentration of fat in the thighs is called *steatomeria*.) A number of Ice Age figurines possess disproportionate buttocks that *may* represent widespread prehistoric steatopygia—or, other commentators argue, may exemplify an artistic convention. Best known of these is the so-called Venus of Willendorf found in Austria, but others include Czechoslovakia's Venus of Dolni Vestonice and France's Venus of Lespugue. Composed respectively of chalk, burned clay, and mammoth ivory, the three date from as long ago as twenty-five thousand years. Only two extant cultures, the natives of the Andaman Islands off the coast of India and the Khoi of South Africa, still exhibit a genetic predisposition for steatopygia. Following upon the usual history of colonial dispersal and extermination, the detribalized culture of South Africa is gradually eroding the distinctive cultural aspects of the Khoi, and the tendency toward steatopygia is fading. Usually, when geographic isolation and restrictions on interbreeding are relaxed, a smaller population's distinct characteristics are diluted by a larger population group. It is the same phenomenon that occurs with any other animal.

Invading Europeans dubbed the Khoi the "Hottentots," and it is by this term that they are still known in much of the West. Indeed, the best-known example of steatopygia is the tragic story of a woman remembered now mostly by her imposed stage name— the Hottentot Venus. She was a Khoi, born in 1789 in what is now South Africa. Her real name, or at least her Afrikaans name, was Saartjie Baartman. In 1810 a doctor named William Dunlop conned Baartman into making the mistake of her life; he told her that she could earn a lot of money displaying her impressive posterior to freak-hungry Europe. Her decision to do so had a tragic impact, both on her own life and on European stereotypes about Africans. Like an animal, she was exhibited in circuses and museums. Learned naturalists examined her. Cartoonists mocked her. One print from the early nineteenth century portrays various onlookers marveling at Baartman's buttocks, one man even standing back and peering through a telescope. Their remarks indicate just how demeaning the experience must have been for the young woman. One observer murmurs, "Ah! How comical is nature," while another exclaims, "Oh! God damn, what roast beef!" Soon, of course, the novelty faded. After almost starving as a prostitute, Saartjie Baartman died in Paris in 1816. Until the mid-1970s, the Museum of Mankind in Paris exhibited her skeleton, and even her brain and genitalia. (Georges Cuvier himself dissected Baartman's labia, which were as hypertrophied as her buttocks, and inevitably he compared the enlarged area to that of apes.) In postcolonial Africa, Baartman became a symbol to the Khoi, and in 1994 the government of South Africa and a committee of the Griqua people, a Khoi subgroup, petitioned France for her return. In 2002 France finally agreed.

Nowadays plastic surgeons prescribe liposuction for steatopygia. In fact, excessive padding in this region is epidemic among Americans and increasingly so elsewhere in the world, as more and more people eat too much junk food and exercise too little. In 2000 an English radio commentator speculated on whether England would become "a nation of fatties" like the United States. And yet many people beside the Khoi consider overdeveloped buttocks erotic. There are even online clubs that cater to this preoccupation, and one enthusiast uses the e-mail pen name "steatopygiahottentot."

The cleft where the buttocks begin to form into two hemispheres—the butt crack famously exhibited by fat plumbers who drop wrenches—was once called the *nock*. The word survives elsewhere, as the name of an arrow's notch to accommodate the bowstring. Some historians speculate that the standard notched-heart outline, so different from the shape of an actual heart, originally portrayed the buttocks, from cleft to curves. Several cartoonists have employed this idea. The theory nicely connects the heart shape with the naked bottoms of Cupid and his simpering colleagues and helps perk up the hoary symbolism of Valentine's Day.

And now for a decidedly brief glance, as porn movies say, "between the cheeks." As everyone in the world knows, the anus is the terminus of the digestive tract, the opening of the culvert that conveys outward the indigestible remains of meals and provides for the emission of gaseous by-products. The anus is a good invention. Consider what our lives would be like without it. Many invertebrates, such as jellyfish, possess relatively simple digestive systems that lack this sort of exit for waste products; they must vomit undigested material, ejecting it via the same orifice through which it entered the body. Most vertebrates have a single cloaca—the source of the name can be seen in the Cloaca Maxima, the main sewer in ancient Rome—through which they expel not only their combined wastes but even their eggs or offspring. As placental mammals, however, we have separate openings, urogenital and anal, for the segregated excretion of liquid and solid waste matter. Born outlaws such as the platypus and the echidna are mammals but are primitive enough to lay their eggs through the same cloaca that transports their waste matter. For this nasty-sounding efficiency, they earned their order the name Monotremata, "one hole." Obviously, however unworthy of our divine aspirations defecation may seem, as a routine it is simpler and less socially objectionable than many of our fellow creatures' alternatives. Regarding the posture we assume to complete the act of digestion, the natural seems to be a squat, but civilization is almost by definition a structured resistance to the natural. In *The Body Has a Head,* Gustav Eckstein contrasts defecation postures of various creatures and ends with a picture of humanity: "Man sits. Man is the thinker."

And yet not even in this pensive attitude can we find the elusive uniqueness for which we yearn. As mentioned elsewhere in this book, Jane Goodall and other primatologists describe watching chimpanzees hold hands, kiss upon greeting and parting, hug, wield tools and weapons, and reach out to touch a companion when afraid. Goodall even tells of watching a female chimp with diarrhea wipe herself with leaves—to be imitated only a moment later by her child, who twice wiped his own clean rear.

The anus and surrounding tissues are a sensitive area, and the hedonistic brain hates to waste a nerve ending that might bestow pleasure. Therefore what began as an exit seems to have long been doubling as an entrance. Because the anus fits snugly around a finger or a penis or other objects, and because the introduction of semen into the rectum does not cause pregnancy, and because the orifice is handy for sex between two men, the so-called back door sees a surprising amount of traffic. Anal sex is one of the common ways to communicate HIV. Both heterosexual and homosexual anal sex has been common throughout history. The act shows up everywhere, from ancient Greek pottery to the graffiti of Keith Haring. In one of his notebooks, Leonardo da Vinci drew an erect penis pointing straight at male buttocks and the well-placed inkblot between them. He also left behind a series of sketches that begins with a portrayal of the five muscles of the anal sphincter (which, in this drawing, resemble a camera aperture), then evolves into the similar five petals of a flower, and winds up as a pentagonal fortress. Inevitably these drawings lead us to Freud—about whom, in this context, we need only note that, while suffering lifelong from constipation, he invented the concept of anal retention.

I can sort of clench my butt, but I don't think it could wiggle a tail.

CALVIN, in *Calvin and Hobbes*, by Bill Watterson

The term *ass backward* means to approach something from the wrong direction or to reverse the order of events. This phrase has a surprisingly ancient lineage. Currently we use the word *preposterous* to mean foolish, senseless, or contrary to nature, but it derives from Latin words meaning preposterior, hind part forward, ass-backward. The phrase *piece of ass* derives not from *ass* so much as from *piece*, as indicated in its synonyms over the centuries, *piece of tail* and *piece of flesh*. In bizarre synecdoche, by the thirteenth century the word *piece* mocked a disreputable member of a company, and by the 1500s, inevitably, it was applied mostly to women. References over the last several hundred years apply the term to women, to their genitalia, and to sexual intercourse with a woman. *Piece of tail* survives, as does just *tail* to refer to the human bottom—as in an American woman overheard telling her child, "Git yer tail in this car before I jerk a knot in it!" Her threat fits with the tradition of such comments as "Kiss my ass!" and a term also popular in the rural South as a synonym for misbehavior, "showing your butt." Apparently the dangerous butt, like the breasts and genitals, is not to be shown, either literally or figuratively. In the 1940s the Catholic Legion of Decency demanded that even the cartoon character Betty Boop must stop showing her garter belt and thighs and the beginning curve of her buttocks. By imitating the dangerous female bottom, those black lines on celluloid were too sexy for the public good.

The etymology of *tail* brings us to one last aspect of the

buttocks—or, rather, to something missing therefrom. For at least a thousand years, the word *tail* has described the usually movable flourish attached to animals, from lizards to horses. Pan, the Arcadian god of flocks and shepherds who became the very spirit of the wild and lent his name to *panic,* naturally had a capric tail to accompany his horns, cloven hooves, and hairy body. So did his Roman equivalent, Faunus. Satyrs and fauns in general possessed tails as evidence of their untamed nature. The very word *satyr* is an imprecise translation of the Hebrew *sa'ir,* which in the Bible refers to both a literal he-goat and a goat demon. The association between tails and wildness still flourishes. Pinocchio's fall into evil ways is first signified by his growth of a donkey's tail. In our contemporary world, human beings have tails only when impersonating other creatures. The list includes the dancers in *Cats* and Michelle Pfeiffer as the fetishistic Catwoman in *Batman Returns.* The late and not lamented icon of male fantasy, the Playboy Bunny, sported a fluffy tail supposedly derived from rabbits but more strongly resembling a white-tailed deer's.

We pin the tail on ourselves for the same reason that we assign bird wings to angels and horse bodies to centaurs—to extend our range, to reembrace the furry creatures that we imagine are playing serf to our lord of the manor, to hug them as totems, to partake of their innocence and power. Most of these urges are unconscious. In other parts of the brain, we are smug about even our taillessness. "According to popular impression," wrote Darwin in *The Descent of Man,* "the absence of a tail is eminently distinctive of man; but as those apes which come nearest to him are destitute of this organ, its disappearance does not relate exclusively to man."

Homoni caudati hic, read some of the blank areas in maps of yore: "Here live men with tails." From the ever gullible Pliny to the openly fictitious Marco Polo, travelers and writers claimed that tribes of humanity bearing tails like animals could still be found—in faraway locales, of course. Just as the incidents in urban legends always happen to a friend of a friend, so do marvels always lurk just beyond the horizon.

TOES WITH SPECIAL PROPERTIES

During his nighttime imprisonment in Dracula's castle, Jonathan Harker opens a window for some fresh air. Already he has discovered that mirrors do not reflect his cold-handed, foul-breathed host, and what he sees outside the window does not reassure him. A movement near the windows of the count's own rooms catches his eye. Harker leans farther out. Dracula's head appears out of a window, and at first Harker is merely curious.

> But my very feelings changed to repulsion and terror when I saw the whole man slowly emerge from the window and begin to crawl down the castle wall over that dreadful abyss, face down with his cloak spreading out around him like great wings. . . . I saw the fingers and toes grasp the corners of the stones, worn clear of the mortar by the stress of years, and thus by using every projection and inequality move downwards with considerable speed, just as a lizard moves along a wall.

What better way to foreshadow a villain's base nature than to describe him performing some action that is characteristic of *other* creatures? It may be incidental to Stoker's purpose that Dracula flaunts his bestially prehensile toes and fingers across the very geometry of domestic order, but the vampire's scurry down the upright walls creates a powerful image.

"Some people are born with bodily parts that possess special properties," wrote Pliny the Elder in the first century c.e. The gullible encyclopedist offered a prime example: "King Pyrrhus' big toe on his right foot cured an inflamed spleen by touch. The story is told that when he was cremated his big toe would not burn along with the rest of his body; it was put in a chest in a temple." Although seldom as useful as the curative digit of Pyrrhus, other toes have left interesting legacies—in literature from Dracula to Borges, in medicine and evolutionary biology, in comic strips and royal scandals, and even in, of all things, the history of spiritualism.

Most of us can pick up a sock with our toes or use a toe to scratch the top of the other foot. But this sort of minor-league toe twisting pales before the discipline of people who actually practice such feats. Houdini's famous dexterity extended even to the farthest outposts of his body. The first series of motion pictures about the great magician and escape artist, a melodramatic serial entitled *The Master Mystery*, featured Houdini performing escape methods that had formerly been hidden from his stage audiences. According to biographer Kenneth Silverman, Houdini's toes were as dexterous as his fingers: "Shackled to a wall, he slips off his shoes and stockings, fishes a ring of keys from the pocket of an unconscious thug, sorts them to select one, and opens the door facing him. All this—snatching the ring, selecting a key, unlocking the door—he does with his toes alone."

Few people, however, school their toes in actions for which they already have handy tools. Some of us allocate our toes only scraps of attention; we simply forget that they are down there until we stub one against a chair leg. "My toes are long and ape-like," says the American essayist Phillip Lopate; "I have very little fellow feeling for them; they are so far away, they may as well belong to someone else." Other people don't feel disengaged at all from their toes. The dissimilar arrangement of nerve endings in the feet makes touching with them different from touching with the hands. Walking barefoot, especially across the ground, conveys a sense of firm reality that is muted and distanced by shoes in much the same way that the windows of a car insulate us from the world we drive through. Still, it is true that we don't see our toes in the mirror when we shave or apply makeup. We don't type with them or glimpse their nails when we raise a coffee mug. You can tell from a glance at many men who wear sandals that their body consciousness fades before it reaches their extremities. At least in the Western world, painting toenails reminds women to groom these outposts. Yet, because the feet are so far from the hands, the act of painting the toenails requires either curling into an upright ball on the floor or standing on one foot and propping up the other.

Our toes have as interesting a history behind them as does any other part of the body. In his 1999 book *The Eternal Trail: A Tracker Looks at Evolution*, the American paleontologist Martin Lockley discusses the toes' position in our bipedal stature. "Books on evolution traditionally talk of the 'rise of animals' from lowly beginnings to higher status," he points out. Lockley explains that this metaphor is literally embodied in the earlier development of *plantigrade* animals, those creatures such as amphibians that plant their entire feet firmly on the ground, followed by the later evolution of *digitigrade* animals, the hoofed mammals and birds and other creatures that walk on their digits. (As discussed in Chapter 8, the word *digit* is from the Latin word for the fingers and toes, which makes counting with digits nicely primitive and appropriate.) This gradual liberation from a ground-hugging posture, adds Lockley, reaches its peak so far in birds and human beings. And yet, "Despite our having free hands, the human foot is plantigrade because both heel and toes contact the ground when we are standing and walking." This action inspires obvious symbolism. In Greek mythology Hercules could not best Antaeus while the giant had his feet planted firmly on the earth. We still respect people who stand their ground.

The digits of our forepaws and hind paws have evolved into quite different tools for their particular jobs. Our numerous features surviving from the adaptations of primate ancestors include fingernails, which are merely flattened claws on our forepaws; the thumb that has moved around to an opposable angle; and the larger, grasping toe that stands out from its fellows. The technical term for the innermost digit on the hind limb of pentadactyl creatures is the *hallux*; in English we call our own version the big or great toe. The English anatomist Frederic Wood Jones wrote a memorable ode to the hallux in his classic 1940s medical text *Structure and Function As Seen in the Foot.*

If there be a human member in which we may justly take pride without laying ourselves open to a charge of self-adulation, that member is the big toe. Philosophers are wont to laud the perfections of our enlarged but simple and but recently emancipated mammalian brain; artists pay homage to certain female bodily contours made for the most part of subcutaneous fat; poets wax eloquent over the simplest biological features when they constitute a part of the human body—but the big toe lacks its champion.

Apparently the big toe genuinely needs a champion. Our dismissive view of this part of the body can be found even in the work of the man who said almost everything better than almost everyone else. Shakespeare's Menenius Agrippa, defender of the harsh policies of Coriolanus, dismisses an outraged citizen as a rabble-rouser by calling him "the great toe" of the assembly.

"I the great toe?" demands the perplexed citizen. "Why the great toe?"

Menenius sarcastically explains that, "being one o' the lowest, basest, poorest" of the rebellious citizens, he nonetheless "go'st foremost."

Despite such slanders, the big toe is an essential, hardworking, much abused, and occasionally attractive member of the body's citizenry. In Bill Watterson's comic strip *Calvin and Hobbes*, the philosophical six-year-old ponders his brief but adventure-filled life and tells his feline best friend that he can list only one disappointment: "I regret I wasn't born with opposable toes." Any child with a tree house will sympathize. All nonhuman primates arrive factory-equipped with a big toe that dramatically opposes the other four: It turns at an angle that facilitates the grasping of tree branches. Our own hallux, while unquestionably the leader of the toes, lacks the ability to turn like a squirrel's. This is why Dracula's scurry down the wall of his castle so unnerves Jonathan Harker.

Our toes may have lost the climbing agility that serves other primates and the occasional vampire, so that they might instead buttress their owner's upright posture, but in exchange they have become the foot soldiers of bipedality. They grasp the ground, anchor us and propel us forward, serve as brakes when necessary. They are so adept at their jobs that they can even perform the comical act of tiptoeing and the elegant but painfully unnatural spectacle of ballet. ("Land like a pussycat," Balanchine advised his dancers.) The toes also serve as balancing base for a woman tottering about in high heels. Our pedestrian big toe is more prominent than that of, say, chimpanzees because it has evolved, like many other areas of the body, to accommodate the changed pressures of upright movement. Different parts of the foot support the body's weight at different moments during the act of walking. The pressure on the foot begins with the heel's first contact with the ground as the leg swings forward; it travels along the side of the foot and then to the ball. The big toe—the foot's last point of contact with the ground—helps steady it and propel it forward. A comparison of a succession of footprints of human beings and chimpanzees reveals not only the dissimilar shapes of their feet but also the different successions of pressures on them. Like our heel bone, the human big toe differs dramatically from that of the chimp partially because it carries weight so differently.

Grasping feet are not limited to primates. The opossum, for example, has a thumblike big toe, and similar adaptations can be seen in a number of creatures, from chameleons to climbing mice. What these animals have in common is life in the trees. "The comparison suggests that this feature is indeed an arboreal adaptation," anthropologist Matt Cartmill has observed, "and that the last common ancestor of the living primates, like most of its descendants, must have been a tree-dwelling creature."

Sometimes our normally well-behaved toes malfunction in a manner that reveals an evolutionary history dating much further back than the innovations of arboreal mammals. Nervous systems began to evolve in segmented animals millions of years earlier. Anyone who has dissected an earthworm in Biology 101 has observed this kind of structure—similar segments divided by walls of muscle. In time each primordial segment not only evolved its own muscles and covering of individual skin but also had its own nerves linked to the proto–spinal cord. Eventually, like nations establishing trade for their mutual benefit, the segments evolved connections that enabled them to work together. Increasing neurological complexity permits ever growing cooperation within the bodily ecosystem, which in turn can lead to even more complex nervous systems and brains. Like other aspects of the body, this process nicely serves as a microcosmic analogy for the development of cooperative social systems.

Few parts of the human body retain evidence of such segmentation. One such record is the abdominal muscles. Another survives in our bodies' retention of a primitive segmented spinal cord, as demonstrated by a related disorder affecting the toes. A movement disorder is defined as any neurological malfunction that causes either abnormal involuntary motion or interferes with automatic motion, at least any not attributed to spasticity or weakness. There is a comparatively rare movement disorder named with prosaic accuracy "painful foot and moving toe syndrome." The neurologist Harold Klawans documents such a case in his book *Strange Behavior: Tales of Evolutionary Neurology*. He describes a patient who sought treatment for ongoing but oddly unfocused pain in her left foot, accompanied by nonstop movement in her toes.

> All her toes were moving in succession, one at a time: up, then down, then back in place as the next toe began to move . . . [T]hey continually followed the same pattern, which resembled a crowd at a tennis game following a match: heads moving in a wave, time after time for as long as I watched. Even when she stood up and put weight on her foot, and even when she walked, the wave kept right on going.

Klawans did not have to search the literature for the poor woman's symptoms; no other disorder causes such behavior. His diagnosis was confirmed by the patient's description of her pain as being both all over her foot and yet not localized in any particular site. One of the early developments in the evolution of unified bodily segments was the concentration of the nervous system in various places along the spine. Think of the two brains of dinosaurs such as *Stegosaurus*, one monitoring the front of the animal and the other the rear. Many creatures retain vestiges of this system. Consider the ability of one dinosaurian descendant, the chicken, to continue scurrying about even after decapitation; the legs do not instantly stop moving when they are disconnected from the brain because they are not operated by the brain. The human spinal cord has two bulges, a cervical enlargement that supplies nerves to the arms and a lumbar enlargement that supplies the legs. Like the human infant walking reflex described earlier in this chapter, the patient's moving toes are controlled not by the brain but by the spinal cord. With her spinal nerves actually causing the trouble, her brain was unable to precisely locate the pain.

Our toes are essential for upright locomotion, but they seldom play a starring role in history. Yet these limber digits certainly get top billing in the story of the Spiritualist movement that swept the United States and Britain in the middle of the nineteenth century and laid the groundwork for today's TV charlatans who prey upon the grieving and credulous. For an era of scientific advances and loosening faith in religious authority, Spiritualism seemed to offer reassuring "evidence" of an afterlife. The movement is generally considered to have begun in 1848 in upstate New York, with pranks by the youthful Margaret Fox and her younger sister Kate. Later Margaret described their historic practical joke as "a most wonderful discovery . . . and all through a desire to do mischief only." They learned that they could snap their big toes and even their ankles to produce surprisingly loud noises that seemed to reverberate from the floor itself. "With control of the muscles of the foot," Margaret later wrote, "the toes may be brought down to the floor without any movement that is perceptible to the eye. The whole foot, in fact, can be made to give rappings by the use only of the muscles below the knee." Neighbors asked questions and were answered with raps. Apparently few asked *why* the dead would choose to communicate through a sort of makeshift Morse code, channeled by children. Spirits, like gods, have always been maddeningly coy.

Soon the girls' older sister bullied them into taking their gifted toes on the road. In adulthood Margaret and Kate continued as spiritualists, holding séances around the world, with their fans including influential celebrities such as Horace Greeley and Elizabeth Barrett Browning. Seldom does a confidence artist admit chicanery, and therefore both believers and skeptics were surprised when, late in life, Margaret Fox offered a full confession. In 1888, before an audience of thousands, she demonstrated the method by which she and her sister faked spirit rapping. The following morning the *New York World* ran a story about what it called Margaret's "preternatural toe joint." A committee of three physicians "unhesitatingly agreed that the sounds were made by the action of the first joint of her large toe." (Thomas Huxley, the Carl Sagan of Victorian naturalists, learned to perform this trick as well as any spiritualist.)

Even though this elderly woman confessed that decades of deceit were troubling her conscience, devout Spiritualists refused to believe her. They insisted that skeptics had forced her to fake a confession in order to further their evil rationalist goals. Four decades later Arthur Conan Doyle praised the Fox sisters in his unintentionally hilarious *History of Spiritualism*. Rather than believe Margaret Fox's confession, the hopelessly credulous Conan Doyle suggested ways in which she might have been unaware of her own supernatural power. Such a task required endowing her toes with special properties worthy of King Pyrrhus. "Let us see . . . ," wrote Conan Doyle, "if any sort of explanation can be found which covers the double fact that what these sisters could do was plainly abnormal, and yet that it was, to some extent at least, under their control." According to one source he quotes, raps are caused by ectoplasm, which Conan Doyle nebulously defines as "a protrusion from the medium's person of a long rod of a substance having certain properties which distinguish it from all other forms of matter. . . . The one assumption is that a centre of psychic force is formed in some part of the body from which the ectoplasm rod is protruded." After an assumption of this magnitude, Conan Doyle's next sentence was inevitable: "Supposing that centre to be in Margaret's foot. . . ."

Adorning these historically influential toes are nails that further demonstrate a kinship with the clawed feet of our country cousins. When Plato classified the human race as featherless bipeds—demonstrating why he is not remembered as a taxonomist—Diogenes supposedly presented a plucked cock to the Academy with the remark, "This is Plato's man." The academicians were forced to extend Plato's definition to include a qualifying clause: "with broad flat nails." Plato needed both parts of the definition. Although our upright stance still distinguishes us, the addition of broad flat nails admits into the definition the fingernailed hands and feet of our fellow primates. We scratch our itches with ancient family heirlooms; distinctive primate nails evolved at least 55 million years ago. Unlike other mammals', primate feet both fore and hind have flattened nails that are more like a shield for the sensitive fingertips than a sword to extend them.

Until recently many evolutionary biologists attributed these features to the distinctive demands of life in trees. Unfortunately, the arboreal theory of primate evolution does not explain why the group developed in this way but so many other tree-dwelling animals did not. Primate nails differ greatly from the curved, gouging, slicing tools that technologically enhance the fingers and toes of other mammals—even those that climb. (In fiction and fable, claws are among the many body parts that we borrow from other creatures and attach to human beings to create or emphasize monstrosity.) If the difference is related to arboreality, why are other creatures perfectly content with their own version? In the common gray squirrel of North America, the digits are basically parallel and the thumb is demoted from versatile tool to grasping utensil. The once flat nails of some primates, such as tamarins and the aye-aye, have anachronistically evolved *back* toward claws.

The Argentine-born fabulist Jorge Luis Borges amusingly commented on the human version of this familiar mammalian attribute. He says of his toes that they are interested only in "turning out toenails—semitransparent, flexible sheets of a horn-like material, as defense against—*whom?* Brutish, distrustful as only they can be, my toes labor ceaselessly at manufacturing that frail armament. . . . By the ninetieth twilit day of their prenatal confinement, my toes had creaked up that extraordinary factory." Borges's description of toenails as hornlike is accurate. Like the nails adorning our redesigned forepaws—and like the horns of cattle—the toenails are composed entirely of keratin, the almost insoluble protein described in Chapter 1. Keratin's durability explains not only why hair clogs sinks but also why nails seem to grow after death: The tissues decay and expose the resilient protein.

Like the navel, the toes seem rather comic. In 1992 England's always scandal-rocked royal family had to endure further ridicule when paparazzi photographed the duchess of York not only cavorting with her noble breasts exposed but also with Texas billionaire John Bryan actually kissing (and apparently sucking) her toes—all while she was still married to Prince Andrew. Had he applied his amorous lips to almost any other body part, Bryan would have looked less clownish.

When they fail us, however, the toes suddenly become as poignant as any other region of the suffering body. Consider an incident that took place late in the polio-ridden life of Franklin Roosevelt. In the last years before his death in 1945, FDR was often accompanied by his cousin Margaret (Daisy) Suckley. Polio had slowly taken away the president's ability to stand; not even his iron grip on a train railing could hold him upright anymore. With touching naïveté, Suckley believed that quack medicines and therapeutic massage would rout the degenerative disease that was moving ahead as fatefully as the Allied forces. She confided optimistically to her diary that after one massage session Roosevelt reported progress: He said he thought that he had been able to slightly wiggle one little toe. How many polio victims despaired at their toes' refusal to obey orders, down there far away from the yearning brain? The body can turn on us in so many ways. As the disease had progressed over decades, FDR's toes had completely lost their most important talent, their ability—normally so reliable, like breathing, that we do not even notice it— to quietly respond to the needs of the body they do so much to support. They could no longer move.

PEDESTRIAN CROSSING

> **pedestrian,** *a.* [L. *pedester,* from *pes, pedis,* the foot.]
> 1. going on foot; walking; also, performed on foot; as, a *pedestrian* journey.
> **pedestrian,** *n.* one who walks or journeys on foot.
>
> *WEBSTER'S UNABRIDGED,* SECOND EDITION

"Although I am an old man, night is generally my time for walking." With this irresistible opening line, Charles Dickens began his novel *The Old Curiosity Shop.* It was 1840, and Dickens was trying to rescue his fledgling periodical, *Master Humphrey's Clock.* Dickens abandoned the perambulations of the elderly narrator at the end of Chapter 3, but the character serves his purpose, which is to introduce Little Nell and her grandfather. He also immortalizes a glimpse of Dickens's favorite pastime: wandering the streets of the city he made his own. Charles Dickens was a passionate walker, by night or by day. "If I couldn't walk fast and far," he confided to a friend during a difficult time in 1854, "I should just explode and perish." In his collection *The Uncommercial Traveller,* Dickens boasted, "My last special feat was turning out of bed at two, after a hard day, pedestrian and otherwise, and walking thirty miles into the country to breakfast." He neglected to mention that this walkathon was in response to a fight with his wife.

A writer whose style depended upon precision and detail, Dickens expressed an important taxonomic distinction: "My walking is of two kinds: one straight on end to a definite goal at a round pace; one, objectless, loitering, and purely vagabond." These two kinds of locomotion describe a great deal of the cultural history of bipedalism. "Walking," writes Rebecca Solnit in Wanderlust, her history of walking, "came from Africa, from evolution, and from necessity, and it went everywhere, usually looking for something." All animals are looking for something—food, shelter, a rival, a mate. Our distinction is that our ancestors, with clever hands completely freed from assisting in locomotion, were able to carry objects with them as they wandered around. Bipedality enabled them to accessorize. Apparently Pliny the Elder was right when he scribbled the epigrammatical line that Isak Dinesen exploited: "Ex Africa semper aliquid novi"—"There is always something new out of Africa." The evidence does indeed indicate that our ancestors struggled upright in this region of the world.

We have come a long way since. Observe the confident bipedality of a contemporary woman on a street in a Western metropolis such as New York. She walks with long strides that convey a sense of dignity and self-respect and power. In most eras of the past, she would not have been permitted to so fully advertise her independence. Even today she is forbidden to do so in many cultures around the world. The right to walk freely may be the first step toward a fair society, as crucial as the right to speak freely. The Chinese custom of foot binding, besides erotically fetishizing a grotesque infantilism in women's feet, also simply kept them unable to walk as freely as men. In his book Flesh and Stone: The Body and the City in Western Civilization, the American cultural historian Richard Sennett explains one of the ways in which our bipedal stance helped stratify a patriarchal society by defining a style of locomotion as particularly masculine: "Erect, equal, purposeful: in Greek, the word *orthos*, or 'upright,' carried the implications of male rectitude." Sennett points out that in Periclean Athens gentlemen were expected to walk with a graceful long stride, pointedly different from the hesitant little steps demanded of women. As a result, of course, men who did not walk in the approved manner were denounced as effeminate. Homer portrayed heroes such as Hector stalking confidently across the battlefield, while the gait of women and goddesses the poet compared to that of timid doves. Such divisions persist. Even the order of White House guests walking into a dining room becomes important among hierarchy-minded herd animals obsessed with our fluctuating rank in the primate troupe. The tireless and irresponsible brain has taken even stride, the primitive action of walking, and twisted it into a tool with which to control others.

Indeed, history has been affected as much by the legs that carried us as by the hands that grasped hammers and the arms that swung them. Our new protohuman legs freed our old primate hands to skillfully grasp weapons that were far more dangerous than the branches hurled by squabbling chimpanzees. Much of history is the record of soldiers walking across the globe, wreaking bloody devastation. Julius Caesar pausing at the Rubicon, Timur Lenk dragging his bad leg while sacking Damascus, Napoleon approaching Moscow—each was leading hordes of walking men. Just as *army* seems to derive from the same root as *arm*, so does *infantry* still refer to soldiers trained and equipped to fight on foot. A great deal of the history of warfare, of scheming old men sending loyal young men into battle, hides in the etymology of this word. *Infantry*'s ancestor, the early French *infanterie*, derived ultimately from *infante*. This term already meant both a boy and a foot soldier, because the Late Latin *infans*, which innocently referred to a youth or young man, had come to mean one who walked into battle. Not until World War I did our constant international strife bequeath us the phrase *walking wounded*, but the reality has accompanied every conflict.

Fortunately, walking can be a tool of peace and justice as

well as of war. In our age of mechanized transport, people walk together in large numbers for only one reason—to deliberately attract attention. Peaceful group walks on behalf of various causes have lent the very act of striding an impressive political resonance, and even describe it with a borrowed military word: *march*. Mohandas Gandhi led a walk, the Salt March, that changed Indian history. In 1930 he defied the British salt monopoly in India with one of his many nonviolent campaigns of civil disobedience. He led thousands of inland natives on a march to the sea, where they insisted upon their right to make their own salt without British taxation or other interference. Martin Luther King Jr., who was profoundly influenced by Gandhi, led the great civil-rights march in Birmingham, Alabama, in 1963, and later the same year walked at the head of another in Washington, D.C. Two years later the march of thousands from Selma to Montgomery, Alabama, drew the attention of the world to the apartheidlike resistance to equality for Americans of African descent. Hundreds of thousands of Americans marched in Washington in 1992 on behalf of reproductive freedom for women. In 1993 almost a million people walked in the National March on Washington for Lesbian, Gay and Bi Equal Rights and Liberation. Many American cities regularly host "Take Back the Night" marches to protest violence against women. In 2002 hundreds of thousands of people around the world marched in protest of President Bush's threatened attack on Iraq. This kind of activity has been going on for some time. In Kensington, Pennsylvania, in 1903, when the American activist Mary Harris "Mother" Jones wanted to force the public to notice the conditions under which children were working in factories, she led them in a march in front of the city hall.

Both King and Gandhi owed a debt to Henry David Thoreau, whose essay "Civil Disobedience" still inspires those who refuse to submit to morally repugnant state policy. In a different essay, Thoreau wrote about the merits of the very action that his civilly disobedient descendants would employ. He wrote "Walking" in 1851 as a lecture, and over the next few years it divided itself into two lectures; they were reassembled and published only after his death. Thoreau had already published the lyrical "A Winter Walk" in 1843, but it was more about where the walk took him and what he saw; "Walking" was about the act itself and its moral implications for the individual and for society. Thoreau trumpets the spiritual virtues of walking in the first sentence of the lecture: "I wish to speak a word for Nature, for absolute freedom and wildness, as contrasted with a freedom and culture merely civil,—to regard man as an inhabitant, or a part and parcel of Nature, rather than a member of society."

Obviously Thoreau was not the first thoughtful walker. Aristotle and his students were called Peripatetics because they discussed philosophy while strolling in the Lyceum in Athens. In *Théorie de la Démarche,* Honoré de Balzac demanded, "Isn't it really quite extraordinary to see that, since man took his first step, no one has asked himself why he walks, how he walks, if he has ever walked, if he could walk better, what he achieves in walking . . . questions that are tied to all the philosophical, psychological, and political systems which preoccupy the world." Few if any writers had asked such a detailed and specific question about the act of walking itself, but many had applauded the corollary virtues of the act. Jean-Jacques Rousseau famously praised the pensive amble in *Rêveries du Promeneur Solitaire.*

William Wordsworth made walking tours part of his manifesto of living and poetry writing. The first recorded usage of *pedestrian* as an adjective comes from one of Wordsworth's letters in 1791. Two years later the noun form appeared in a volume entitled *The Observant Pedestrian.* Robin Jarvis points out in his book *Romantic Writing and Pedestrian Travel* that the era of the Romantic poets saw an impressive flowering of the connection between walking and nature appreciation. Coleridge, for example, not only wrote about walking and while walking; he employed metaphors reminiscent of his own ambulatory preferences, such as eschewing "the beaten high-Road of diction" for the "soft, green pathless field of Novelty." The poet's uneven gait when hiking reminded Hazlitt and other commentators of his conversational style.

Jane Austen's characters take walks in order to converse, think, or commune, although for some readers these rambles are sadly lacking in the sort of descriptive specifics that bring alive the vivid treks of Dickens's characters. Elizabeth Bennett in *Pride and Prejudice* walks for miles to help her ailing sister; Catherine Morland and the Tilney walking party in *Northanger Abbey* amusingly discuss aesthetics as they stroll. In *Walking, Literature, and English Culture,* Anne D. Wallace summarizes the significance of perambulation preferences in Austen's books:

> . . . a taste for walking and respect for other pedestrians become signs of the virtues Austen ascribes to the best of the English landed gentry and freeholders. Gentlemen and gentlewomen are distinguished as readily by their willingness, even their desire, to walk as by their common sense, independence of opinion, unostentatious economy, country living, and what Mr Knightley calls "English delicacy towards the feelings of others!"

Austen expresses the uncertain social footing of the newly mobile bourgeoisie partially by contrasting the attitudes of walkers and riders, in both town and country. The titular heroine of *Emma* has succumbed to the ideas of her age:

> Mr Knightley keeping no horses, having little spare money and a great deal of health, activity, and independence, was too apt, in Emma's opinion, to get about as he could, and not use his carriage so often as became the owner of Downwell Abbey.

Emma was one of many nineteenth-century citizens who found themselves intoxicated by the possibilities of vehicular travel—to the detriment of walking as a leisure activity. Later in the century, the American nature writer John Burroughs complained about the current state of walking in his essay "The Exhilarations of the Road." Like Thoreau, he applauded bipedality's natural legacy. "Man takes root at his feet," he insisted, "and at best he is no more than a potted plant in his house or carriage till he has established communication with the soil by the loving and magnetic touch of his soles to it." It is rather surprising that Burroughs referred to our shod feet rather than suggesting that, like Antaeus in Greek mythology, we might maintain our connection with the Earth only while touching it with our bare feet. Burroughs had noticed "that at our fashionable watering-places nobody walks. . . ." Also like Thoreau, he moralizes: "We have fallen from that state of grace which capacity to enjoy a walk implies. . . . Indeed, I think it would be tantamount to an astonishing revival of religion if the people would all walk to church on Sunday and walk home again." Then Burroughs returns to his preoccupation with shoes. He worries that "the American is becoming disqualified for the manly art of walking by a falling off in the size of his foot. . . . A small, trim foot, well booted or gaitered, is the national vanity."

In 1951, exactly a century after Thoreau's lecture, so many things had changed in the matter of transportation that Ray Bradbury published a short story envisioning how walkers might suffer after yet *another* century. The sole character in "The Pedestrian," Leonard Mead, during his walks "would stand upon the corner of an intersection and peer down long moonlit avenues of sidewalk in four directions, deciding which way to go, but it really made no difference; he was alone in this world of 2053 A.D., or as good as alone." Mead never encounters anyone on his nocturnal rambles. People remain in their homes at night, transfixed by passive electronic entertainment. By the middle of the twentieth century, when Bradbury wrote the story, the television was already becoming the new hearth fire, and families were building their evenings around it. Finally an automated police car stops and interrogates the pedestrian:

"What are you doing out?"
"Walking," said Leonard Mead.
"Walking!"
"Just walking."

The police car takes Mead to the Psychiatric Center for Research on Regressive Tendencies. This may be heavy-handed satire, but it was prophetic. Precisely at the time of Bradbury's story, bureaucrats began using the term *pedestrian precinct* to designate an area daringly intended for walkers rather than for drivers. By the early 1960s, they had coined the hideous word *pedestrianize* to describe areas undergoing attempts to restore a sense of community after a half century of automobile-oriented fragmentation. "The world has become inaccessible," wrote Ivan Illich, "because we drive there." Countless sociologists and city planners have documented the relationship between walking and community. As Rebecca Solnit says in *Wanderlust,* "the history of suburbia is the history of fragmentation." Carl Sagan caught the attention of college audiences on his lecture tours by pointing out that, if aliens are watching us from outer space, they must be assuming that automobiles are the dominant life-form; we service them, maintain them, fuel them, house them, and build roads for their convenience rather than our own. In the 1930s Lord Dewar, the epigrammatical British whiskey maker, summed up pedestrians' relationship to vehicles in the century after Jane Austen when he remarked that there were already "only two classes of pedestrians in these days of reckless motor traffic—the quick, and the dead."

An unlikely consequence of our antipedestrian society is now an essential item in gyms throughout the Western world, and it is also common in homes: the treadmill, a stationary track on which our legs can exercise their ancestral talents without actually going anywhere. This device—worthy of Kafka or Beckett or the Red Queen—was originally invented in 1818, by an

Englishman named William Cubitt. It was designed to break down what Cubitt called the "obstinate spirit" of prison inmates with its insufferable monotony. The original, however, produced energy because the inmates walked on it; in our time it absorbs the energy produced by the walkers. To Cubitt's surprise it turned out to have beneficial health effects for the prisoners, just as it does nowadays for those imprisoned indoors by an automobile society.

Ray Bradbury's story came true—had always been true—in other ways, too. A California man named Edward Larson, who liked to walk in his neighborhood at night, was arrested fifteen times by police officers before he challenged a certain law in 1983. The restriction made it illegal in his state to walk at night unless the pedestrian was willing to present acceptable identification when approached by a police officer, whether or not the officer could name "probable cause" for accosting the walker. Larson's experience demonstrates one of the many ways that walkers can literally step outside the boundaries of the majority. He was an African American male outdoors at night, and therefore white society labeled him as suspect and potentially dangerous. Groups in power often restrict where groups that are less powerful (and therefore potentially dangerous to the status quo) may walk.

Stepping outside the lines can be even more difficult for women. "The geography of race and gender are different," says Rebecca Solnit, "for a racial group may monopolize a whole region, while gender compartmentalizes in local ways." A young New Yorker of too low a social class and too feminine a gender was arrested in 1895 because she was walking home from her aunt's house and asked two men for directions. She suffered an examination to prove her virginity before she was released. Almost a century later, in the early 1980s, the media frenzy surrounding the trial of Peter Sutcliffe, the so-called Yorkshire Ripper, included many attempts to somehow blame the eleven dead and six wounded women for their fate. Both law-enforcement officials and news media were quick to imply—and sometimes explicitly state—that not only the prostitutes but also the students and housewives who died *deserved* to be punished for walking late at night unaccompanied by a man. Under the Taliban in Afghanistan, in Saudi Arabia, and in many other places around the globe, now and throughout history, women have been controlled first by restricting where they may walk without a male companion.

The concept of a walk still shows up everywhere, and so does the word itself. If you walk on eggshells, someone may walk all over you. Certain businesses welcome walk-ins, but strikers stage walkouts. Novice thespians settle for walk-ons. In sports a *walk* usually refers to a demotion from the normal gait of players. In baseball you walk after four balls; in basketball, if you walk, you are moving while holding the ball; in cricket you leave the field. Because we need to communicate during our perambulations, soon after the invention of radio we had a portable version called in military babyspeak a *walkie-talkie,* but not until 1981 did Sony patent a portable radio and music player called the Walkman. Modern language manufacturers have surgically removed the last syllable of the word *marathon,* misidentified it as a suffix, and grafted it onto other activities. Now we have such variations as *telethon,* but *walkathon* was first. It emerged during the Depression and originally referred to a competitive activity. It has since come to mean a fund-raiser in which participants have been promised money in proportion to the distance that they walk, no matter in which order they cross the finish line.

Competing or collaborating, we sometimes need help walking. "What creature," the Sphinx demanded of Oedipus, "walks in the morning on four feet, at noon upon two, and at evening upon three?" Wise Oedipus saw through the double riddle and replied, "Man." Human beings are quadrupedal when they crawl in infancy, bipedal when they walk erect in adulthood, and tripedal when they lean upon a cane in old age. In the United States, *walker* is a term for a rolling contraption that helps an infant prepare for bipedality. In line with the Sphinx's riddle, the

word also names a similar invention that assists those who—
through accident, illness, or the subtractions of age—have returned to the infant's dependent state. Although the English
word *ambulance* refers to a vehicle that conveys those who cannot transport themselves, the same word in French named an entire field hospital, which in turn got its name from a word for
itinerant, which had evolved from the Latin word meaning to
walk—*ambulare*. Today hospitals label a patient not confined to
bed *ambulatory*, and a vehicle with which a parent or baby-sitter
can transport a not-fully-bipedal child is a *perambulator*, or *pram*.

We naked apes naturally explore the world on our own two
feet. Human beings have walked from Africa to Tierra del Fuego,
across the Bering Strait and the Isthmus of Panama, over the Hi-
malayas and the Andes. We walked barefoot and shod. Formerly
we walked bearing mastodon flesh and quivers of arrows; now we
carry briefcases and cellular phones. Like our ancestors, we traipse,
plod, meander, hike, trudge, saunter, march, stomp, promenade,
stroll, amble, stride, mosey, and occasionally slog. We toe the
line, step lively, drag our feet, and hoof it. The same clever adaptation that carried *Australopithecus* through the falling ash at Laetoli stands on pointe in Stravinsky's *Firebird*. Yes, the action that
Samuel Johnson called walking on our hinder legs has its unfor-
tunate side effects, but it has served us well all over the globe.
Actually, we have already walked *beyond* the globe. "By far the
easiest and most natural way to move on the surface of the
moon," observed astronaut Neil Armstrong, "is to put one foot in
front of the other." Upright walking will continue to transport
our ancient vertebrate form, even far away from our native solar
system. Our descendants will be aiming their mammalian faces at
sights and sounds, and carrying tools in their primate hands,
when they walk through the exotic landscapes of distant worlds.

ACKNOWLEDGMENTS

From New Orleans to Nova Scotia, Laura Patterson has accompanied and assisted the evolution of *Adam's Navel*. Literally from the day that we met, she recommended and provided sources and responded with insight and wit to my ideas and my expression of them. Her brain, talents, and companionship enrich my writing and my life. Thanks, Dr. Patterson.

My agent, Heide Lange, believed in the idea for this book from the first, guided the evolution of the proposal, and agented the U.S. deal while on holiday. She has been the single most influential person in my writing career. Thanks also to other fine people at Sanford J. Greenburger Associates, Inc.: Heide's terrific assistant, Esther Sung, for work on this and other projects, and Peter McGuigan, for handling the foreign sales. (And once again I propose a toast to the memory of Diane Cleaver.)

Throughout the process of writing *Adam's Navel*, I have profited from the guidance and encouragement of two publishers. Stefan McGrath, editorial director at Penguin UK, started this ball rolling by suggesting that I might want to write a book for Penguin. Martin Toseland skillfully guided the book through revision and publication in England. Molly Stern, my editor at Viking, encouraged, advised, and applied invaluable line-by-line attention to the first draft. This book is infinitely better because of her critique. Her assistant, Jennifer Jackson, helped in countless ways. My thanks also to Nancy Resnick, who designed the interior layout; Juliette Borda, who painted the jacket and interior illustrations; and Maggie Payette, who designed the jacket. Thanks also to copyeditor Maureen Sugden and production editor Bruce Giffords.

Richard Milner, historian of evolution, commented upon the entire manuscript, recommended and provided sources, connected me with scholars, and led jaunts through snowy Manhattan. William Sheehan, psychiatrist and science historian, drew upon his vast knowledge about the body to contribute and advise on matters medical and historical, as he has done since I first conceived the idea for this book. Martin Gardner sent me articles and citations and jokes about the body. Primatologist Frans de Waal critiqued some sections of the book that address evolutionary similarities between ourselves and our hairier cousins. James Moore, historian and Darwin biographer, caught some errors. Cultural critic Laura Patterson made countless suggestions, on topics from the natural history of hair to the cultural history of walking. Mark Wait, dean of the Blair School of Music at Vanderbilt University, critiqued various parts of the manuscript and suggested that Robert Schumann's troubles offer an excellent example of damage resulting from the hands' neurological complexity. This book and this author have profited from the writings, gifts of sources, and sheer intellectual energy of Jonathan Miller (recently knighted Sir Jonathan), over meals and during rambles in both the United States and England. Books and articles by several of the aforementioned also contributed enormously, and many are mentioned in the bibliography.

I wouldn't know what questions to ask these scholars without the resources and fine staff of the Jean and Alexander Heard Library at Vanderbilt University. I want to thank especially Dale Manning, Daisy Whitten, Dewey James, and David Carpenter of the Central Library; Jon Erickson and Debra Stephens of the Sarah Shannon Stevenson Science and Engineering Library; and Mary Teloh, of the Special Collections division of the Annette and Irwin Eskind Biomedical Library. (And thanks, Mary, for moving the skull.) The absurdly early death of Chris Germino, Vanderbilt librarian and good human being, demonstrates the idiocy of the universe. I must applaud the resources and endlessly helpful staff of the Nashville Public Library. Andre Norton cordially provided access to the High Hallack Genre Writers Library and gave me source books. My thanks also to the tranquil little oasis of the campus and the Virginia Davis Laskey Library at the Scarritt-Bennett Center in Nashville. Longtime gratitude to the Art Circle Public Library in Crossville, Tennessee, where I learned the quiet joy of prowling, which only later I started calling "research"; I especially thank the director, Debbie Hall, and the library's former director, Marilyn Foster. I also want to express my appreciation, long after her death, to Mary Belle Rogers, the librarian of Homestead Elementary School in Crossville, Tennessee, when I was there in the 1960s. When I consider how welcome and comfortable I feel in libraries, I remember her voice, smile, and enthusiasm for books.

For providing specific insights for this book, and also for furthering my education and enriching my life, I welcome this opportunity to thank several fine museums. I examined bodies and images of bodies in Boston at the Museum of Fine Arts; in Cambridge, Massachusetts, at the Harvard Museum of Comparative Zoology; in New York City at the Metropolitan Museum of Art, the Frick Collection, and the American Museum of Natural History; in London at the National Gallery, the British Museum, the Natural History Museum, and the Victoria and Albert Museum; in Paris at the Musée National du Moyen Age (the Musée de Cluny); and in Heidelberg at the wonderful little Volkerkundemuseum.

Many people assisted me while I visited these places and others during the writing of this book. Martin Blaser, chair of the School of Medicine at NYU, and Ronna Wineberg-Blaser provided hospitality in Greenwich Village and facilitated my visit to Paris. Agnès Labigne of the Institut Pasteur drove me around Paris. Captain Jeannine Smith provided her lovely apartment in Heidelberg and guidance through Germany, Austria, Italy, and Switzerland. Havi Smith and Armand LePage gave meals and advice in Germany. Andrea McQuay and Ray Conners put me up in Boston. Jack Pennock lent us his house on the coast of Prince Edward Island. Marilyn Matasick lent me her condo on Kiawah Island, South Carolina. My lifelong pal Jeff Hood repeatedly provided his house in Tennessee as a country refuge. Often I worked on this book in my two favorite Nashville haunts, Calypso Café and the coffee shop Fido, and I hereby express my appreciation to the fine staff of each. I especially thank Doug and Brandi, and to Robert I must say, "Thank you, sir."

I am fortunate to have several friends whose careful readings improved many parts of this book. They include Pamela Burdett, Jon Erickson, Laurie Parker, Sally Schloss, and Mark Wait. Michele Flynn and Annabeth Headrick read and commented upon a section apiece. As always, Jim Young recommended and lent many sources. Ed Penney shared his stories of jazz giants in Boston. Dennis Wile took the jacket and publicity photographs and immortalized a funny and tipsy afternoon in London; he also lent books. Vicki Jones ran my life back home while I was out of town for weeks at a time, and killed only one plant.

Many other people contributed in various ways, and I thank them all: Denny Adcock, Rebecca Bain and WPLN public radio in Nashville; Roger Bishop; Sallie Bissell; Alan Bostick and the *Tennessean*; Kae Follis Cheatham; Ellen Chodosh and Oxford University Press; Beth Conklin and *Medical Anthropology Quarterly*; Tamara Crabtree and Ingram Book Company; John Egerton; Cathy Fenner; Phyllis Gobbel; Martha Whitmore

Hickman; Carolyn Householder and Charles May at Bodacious Books; Dolly Kelly, Ginger Knight, and Erin Coston at Davis-Kidd Bookstores; Jordan Lee; Amy Lynch; Galyn Martin and Katie Hoy with the Southern Festival of Books; Jonathan Marx and Bruce Dobie at the *Nashville Scene*; Sandy Matasick; Sue McClure; Madeena Spray Nolan; Randy O'Brien at WMOT public radio at Middle Tennessee State University; Cathie Pelletier; Kathleen Penney; Sidney Perkowitz at Emory University; Rob Simbeck; Bruce Tierney; Ron Watson; Alana White; F. Clark Williams; and Larry and Saralee Woods at BookMan/BookWoman.

My family provided endless encouragement. Joseph Randall, Yow Greg Norris, and Janet Derrick, cousins and friends, provided the skewed perspectives that have entertained me for decades. Her Honor Mary Ann Yow McNabb kept me connected with the family. My cousin Helen Derrick has been an inspiration ever since she bought me copies of *Writer's Digest* when I was a teenager. Lifelong gratitude goes to my brother, David Sims, whose intellectual interests and mine so satisfyingly intersect nowadays. My mother, Ruby Norris Sims, introduced me to books by holding me on her lap and reading to me, and here we are four decades later, still prowling used bookstores together. Thanks, gang.

SELECTED BIBLIOGRAPHY AND FURTHER READING

The world is a never-ending cross-reference.

CEES NOOTEBOOM

The following bibliography is by no means an exhaustive roll call of the hundreds of books, articles, and reference works that I consulted while writing *Adam's Navel*. Instead it is an edited list of major sources in which you will find further information, related entertainment, and the context for quotations. I list more books than articles because the former are more readily accessible to the general reader. In an alphabetical list of this sort, it is impossible to emphasize which works were the most important; a better sense of such indebtedness may be gained from the text.

Abrahams, Mark, quoted. "Birds, barking, beer and bellies." *BBC Online,* 4 Oct 2002.

Achenbach, Joel. *Captured by Aliens: The Search for Life and Truth in a Very Large Universe.* New York: Simon & Schuster, 1999.

Achtemeier, Paul J., et al. *Harper Collins Bible Dictionary.* San Francisco, Calif.: HarperCollins, revised edition 1996.

Angier, Natalie. *Woman: An Intimate Geography.* Boston: Houghton Mifflin, 1999.

Armstrong, Louis. *Louis Armstrong in His Own Words: Selected Writings,* edited by Thomas Brothers. Oxford: Oxford U. Pr., 1999.

Ayto, John. *20th Century Words.* Oxford: Oxford U. Pr., 1999.

Ball, Philip. "These Feet Were Made for Walking." *Nature* online, 25 July 2001.

Barrett, Leonard E. *The Rastafarians: Sounds of Cultural Dissonance*. Boston: Beacon Press, 1977.

Beauvoir, Simone de. *The Second Sex*, translated and edited by H. M. Parshley. New York: Knopf, 1993.

Bell, Robert E. *Dictionary of Classical Mythology: Symbols, Attributes & Associations*. Santa Barbara: ABC-Clio, 1982.

Belluck, Pam. "The Night Before: A Mundane Itinerary on the Eve of Terror." *New York Times* (Reuters), 5 October 2001.

Bentley, G. E., Jr. *The Stranger from Paradise: A Biography of William Blake*. New Haven: Yale U. Pr., 2001.

Bergman, Ingmar. *The Seventh Seal (Det Sjunde inseglet)*, filmscript translated by Lars Malmström and David Kushner. London: Lorrimer, 1984.

Bergreen, Laurence. *Louis Armstrong: An Extravagant Life*. New York: Broadway, 1997.

Biddle, Wayne. *A Field Guide to Germs*. New York: Henry Holt, 1995.

Blitman, Joe. Interview with author, 1999.

Bondeson, Jan. *A Cabinet of Medical Curiosities: A Compendium of the Odd, the Bizarre, and the Unexpected*. Ithaca, N.Y.: Cornell U. Pr., 1997.

Bordo, Susan. *The Male Body: A New Look at Men in Public and in Private*. New York: Farrar, Straus and Giroux, 1999.

Borges, Jorge Luis. *Collected Fictions*, translated by Andrew Hurley. New York: Viking, 1998.

———. "The Creation and P. H. Gosse," in *Other Inquisitions 1937–1952*, translated by Ruth L. C. Simms. Austin, Tex.: U. of Texas Pr., 1964.

Boston Women's Health Book Collective. *The New Our Bodies, Ourselves: A Book by and for Women*. New York: Simon & Schuster, 1992.

Boswell, James. *Boswell's London Journal 1762–1763*, edited by Frederick A. Pottle. New York: McGraw-Hill, 1950.

Bradbury, Ray. "The Pedestrian," in *The Golden Apples of the Sun*. New York: Doubleday, 1953.

Bramly, Serge. *Leonardo: Discovering the Life of Leonardo da Vinci*. New York: HarperCollins/Edward Burlingame, 1991.

Bridie, James. "The Umbilicus," reprinted in Gordon, Richard, editor, *The Literary Companion to Medicine*. New York: St. Martin's, 1993.

Broekman, Marcel. *The Complete Encyclopedia of Practical Palmistry*. Englewood Cliffs, N.J.: Prentice-Hall, 1972.

Brown, Slater. *The Heyday of Spiritualism*. New York: Hawthorn Books, 1970.

Browne, Janet. "Darwin and the Expression of the Emotions," in Kohn, David, editor, *The Darwinian Heritage*. Princeton, N.J.: Princeton U. Pr., 1985.

Bruce, Vicki, and Andy Young. *In the Eye of the Beholder: The Science of Face Perception.* Oxford: Oxford U. Pr., 1998.

Burenholt, Göran, general editor. *The First Humans: Human Origins and History to 10,000 b.c.* San Francisco: HarperSanFrancisco, 1993.

Byrd, Ayana D., and Lori L. Tharps. *Hair Story: Untangling the Roots of Black Hair in America.* New York: St. Martin's, 2001.

Caird, Rod. *Ape Man: The Story of Human Evolution.* New York: Macmillan, 1994.

Calvino, Italo. *Mr. Palomar,* translated by William Weaver. New York: Harcourt Brace Jovanovich, 1985.

Campbell, Horace. *Rasta and Resistance: From Marcus Garvey to Walter Rodney.* Trenton, N.J.: Africa World Press, 1987.

Caplan, Frank, general editor. *The First Twelve Months of Life: Your Baby's Growth Month by Month.* New York: Grosset & Dunlap, 1974.

Cartmill, Matt. "Non-human Primates," in *The Cambridge Encyclopedia of Human Evolution,* edited by Jones et al., q.v.

Carus, Paul. *The Story of Samson and Its Place in the Religious Development of Mankind.* Chicago: Open Court, 1907.

Chamberlain, Andrew T., and Michael Parker Pearson. *Earthly Remains: The History and Science of Preserved Human Bodies.* Oxford: Oxford U. Pr., 2001.

Close, Frank. *Lucifer's Legacy: The Meaning of Asymmetry.* Oxford: Oxford U. Pr., 2000.

Cocteau, Jean. *Past Tense: The Cocteau Diaries, Volume One,* translated by Richard Howard, annotated by Pierre Chanel. New York: Harcourt Brace Jovanovich, 1987.

Coleman, M. J., and A. Ruina. "An uncontrolled walking toy that cannot stand still." *Physical Review Letters,* 80, 3658–61 (1998).

Connolly, Peter, and Hazel Dodge. *The Ancient City: Life in Classical Athens and Rome.* Oxford: Oxford U. Pr., 1998.

Cooper, J. C. *An Illustrated Encyclopedia of Traditional Symbols.* London: Thames & Hudson, 1978.

Corson, Richard. *Fashions in Hair: The First Five Thousand Years.* London: Peter Owen, 1971.

Cosio, Robyn, with Cynthia Robins. *The Eyebrow.* New York: ReganBooks/HarperCollins, 2000.

Cunningham, John D. *Human Biology.* New York: Harper & Row, 1983.

Darwin, Charles. *The Collected Papers of Charles Darwin,* edited by Paul H. Barrett. Chicago: U. of Chicago Pr., 1980. Two volumes in one.

————. *The Descent of Man, and Selection in Relation to Sex.* London: John Murray, 1871.

————. *The Expression of the Emotions in Man and Animals,* edited and annotated by Paul Ekman. New York: HarperCollins, 1998.

————. *On the Origin of Species by Means of Natural Selection.* London: Murray, 1859.

Davenport, Guy. *A Balthus Notebook.* New York: Ecco, 1989.

Davidson, Keay. *Carl Sagan: A Life.* New York: Wiley, 1999.

da Vinci, Leonardo. *The Notebooks of Leonardo da Vinci,* translated and annotated by Edward MacCurdy. New York: Braziller, 1955.

Dawkins, Richard. *The Blind Watchmaker: Why the Evidence of Evolution Reveals a Universe without Design.* New York: Norton, 1986.

————. *Climbing Mount Improbable.* New York: Norton, 1996.

de Grazia, Edward. *Girls Lean Back Everywhere: The Law of Obscenity and the Assault on Genius.* New York: Random House, 1992.

de Sola Pinto, Vivian, editor. *William Blake.* London: Batsford, 1965.

de Waal, Frans. *The Ape and the Sushi Master: Cultural Reflections by a Primatologist.* New York: Basic, 2001.

Dement, William C., and Christopher Vaughan. *The Promise of Sleep: A Pioneer in Sleep Medicine Explores the Vital Connection Between Health, Happiness, and a Good Night's Sleep.* New York: Dell, 1999.

Dickens, Charles. *The Old Curiosity Shop.* New York: Heritage Press, 1941. Also see introduction by John T. Winterich.

Doyle, Arthur Conan. *The History of Spiritualism.* London: Cassell, 1926.

Eckstein, Gustav. *The Body Has a Head.* New York: Harper & Row, 1970.

Edwards, Anne. *The DeMilles: An American Family.* New York: Abrams, 1988.

Ekman, Paul, editor. *Darwin and Facial Expression: A Century of Research in Review.* New York and London: Academic Press/Harcourt Brace Jovanovich, 1973.

Elias, Norbert. *The History of Manners,* vol. 1, *The Civilizing Process,* translated by Edmund Jephcott. New York: Pantheon, 1982.

Eliot, Lise. *What's Going On in There? How the Brain and Mind Develop in the First Five Years of Life.* New York: Bantam, 1999.

Ensler, Eve. *The Vagina Monologues.* New York: Villard, 1998.

Etcoff, Nancy. *Survival of the Prettiest: The Science of Beauty.* New York: Random House, 1999.

Felman, Shoshana. *What Does a Woman Want? Reading and Sexual Difference.* Baltimore: Johns Hopkins U. Pr., 1993.

Fenton, Sasha, and Malcolm Wright. *Palmistry: How to Discover Sex, Love, and Happiness.* New York: Crescent, 1996.

Fisher, Angela. *Africa Adorned.* New York: Abrams, 1984.

Fisher, Helen. *Anatomy of Love: A Natural History of Monogamy, Adultery, and Divorce.* London: Simon & Schuster, 1992.

Fleagle, John C. "Primate Locomotion and Posture," in *The Cambridge Encyclopedia of Human Evolution,* edited by Jones et al., q.v.

Fliedl, Gottfried. *Gustav Klimt, 1862–1918: The World in Female Form.* Köln: Taschen, 1998.

Forrest, D. W. *Francis Galton: The Life and Work of a Victorian Genius.* New York: Taplinger, 1974.

Fox, Margaret. Her confession quoted, *New York Herald,* 24 September 1888; and in "Spiritualism Exposed: Margaret Fox Kane Confesses Fraud," *New York World,* 21 October 1888.

Fraser, George MacDonald. *The Hollywood History of the World.* London: Harvill, 1996, revised edition.

Freud, Sigmund. *Three Essays on the Theory of Sexuality,* translated by James Strachey. New York: Basic, 1975.

Friedman, David M. *A Mind of Its Own: A Cultural History of the Penis.* New York: Free Press, 2001.

Galdikas, Biruté. *Reflections of Eden: My Years with the Orangutans of Borneo.* Boston: Little, Brown, 1995.

Gardner, Martin. "Cal Thomas on the Big Bang and Forrest Mims," and "Robert Gentry's Tiny Mystery," in *On the Wild Side.* Buffalo, N.Y.: Prometheus, 1992. Both originally *Skeptical Inquirer* columns.

———. *Did Adam and Eve Have Navels?* New York: Norton, 2000.

———. *The New Age: Notes of a Fringe Watcher.* Buffalo, N.Y.: Prometheus, 1988.

———. *The New Ambidextrous Universe: Symmetry and Asymmetry from Mirror Reflections to Superstrings.* New York: Freeman, third revised edition, 1990.

Gettings, Fred. *The Book of the Hand: An Illustrated History of Palmistry.* London: Paul Hamlyn, 1965.

Gillman, Susan, and Forrest G. Robinson, editors. *Mark Twain's Puddn'head Wilson: Race, Conflict, Culture.* Durham, N.C.: Duke U. Pr., 1990.

Glaser, Gabrielle. *The Nose: A Profile of Sex, Beauty, and Survival.* New York: Atria/Simon & Schuster, 2002.

Goodall, Jane. *The Chimpanzees of Gombe: Patterns of Behavior.* Cambridge, Mass.: Belknap/Harvard, 1986.

————. *In the Shadow of Man*. Boston: Houghton Mifflin, 1971; revised edition, 1988.

Gordon, Richard. *An Alarming History of Famous and Difficult Patients*. New York: St. Martin's, 1997.

————, editor. *The Literary Companion to Medicine*. New York: St. Martin's, 1993.

Gosse, Philip Henry. *Omphalos: An Attempt to Untie the Geological Knot*. London: John Van Voorst, 1857.

Gould, George M. *Anomalies and Curiosities of Medicine: Being an Encyclopedic Collection of Rare and Extraordinary Cases. . . .* Philadelphia: Saunders, 1900.

Gould, Stephen Jay. *The Flamingo's Smile: Reflections in Natural History*. New York: Norton, 1985.

————, general editor. *The Book of Life: An Illustrated History of the Evolution of Life on Earth*. New York: Norton, 2001, revised edition.

Green, Peter. *Ancient Greece: An Illustrated History*. New York: Thames & Hudson, 1979.

Gruber, Howard E. *Darwin on Man: A Psychological Study of Scientific Creativity*. Published in same volume with Barrett, Paul H., editor. *Darwin's Early and Unpublished Notebooks*. New York: Dutton, 1974.

Haiken, Elizabeth. *Venus Envy: A History of Cosmetic Surgery*. Baltimore: Johns Hopkins U. Pr., 1997.

Haskins, Jim, and Joann Biondi. *From Afar to Zulu: A Dictionary of African Cultures*. New York: Walker, 1995.

Hillman, David, and Carla Mazzio, editors. *The Body in Parts: Fantasies of Corporeality in Early Modern Europe*. New York and London: Routledge, 1997.

Hole, John W., Jr. *Human Anatomy and Physiology*. Dubuque, Iowa: Brown, third edition, 1984.

Hood, Jeffrey. E-mail to author, 2001.

Houdini, Harry. *A Magician Among the Spirits*. New York: Harper, 1924.

Hrdy, Sarah Blaffer. *Mother Nature: Maternal Instincts and How They Shape the Human Species*. New York: Ballantine, 2000.

Hunt, Kevin D. "The pastural feeding hypothesis: An ecological model for the evolution of bipedalism." *South African Journal of Science*, 92 (1996):77–90.

Jahme, Carole. *Beauty and the Beasts: Woman, Ape and Evolution*. New York: Soho, 2000.

Jarvis, Robin. *Romantic Writing and Pedestrian Travel*. London: Macmillan, 1997.

Johanson, Donald, and Maitland Edey. *Lucy: The Beginnings of Humankind.* New York: Simon & Schuster, 1981.

Jolly, Alison. *The Evolution of Primate Behavior.* New York: Macmillan, 1972.

———. *Lucy's Legacy: Sex and Intelligence in Human Evolution.* Cambridge, Mass.: Harvard U. Pr., 1999.

Jones, Dylan. *Haircults: Fifty Years of Styles and Cuts.* London: Thames & Hudson, 1990.

Jones, Frederic Wood. *Principles of Anatomy as Seen in the Hand.* London: Baillière, Tindall and Cox, 1919.

———. *Structure and Function As Seen in the Foot.* London: Baillière, Tindall and Cox, 1944.

Jones, Steven; Robert Martin; and David Pilbeam, editors, with Sarah Bunney, executive editor. *The Cambridge Encyclopedia of Human Evolution.* Cambridge: Cambridge U. Pr., 1992.

Joyce, James. *Ulysses: The Corrected Text,* edited by Hans Walter Gabler with Wolfhard Steppe and Claus Melchior. New York: Random House, 1986.

Kasson, John F. *Houdini, Tarzan, and the Perfect Man: The White Male Body and the Challenge of Modernity.* New York: Hill and Wang, 2001.

Kemp, Martin, and Marina Wallace. *Spectacular Bodies: The Art and Science of the Human Body from Leonardo to Now.* Berkeley: U. of California Pr./Hayward Gallery, 2000.

Kimmel, Michael. *Manhood in America: A Cultural History.* New York: Free Press, 1996.

Kingdon, Jonathan. "Facial Patterns as Signals and Masks," in *The Cambridge Encyclopedia of Human Evolution,* edited by Jones et al., q.v.

Klawans, Harold. *Strange Behavior: Tales of Evolutionary Neurology.* New York: Norton, 2001.

Knappert, Jan. *African Mythology: An Encyclopedia of Myth and Legend.* London: Diamond, 1995.

———. *Pacific Mythology: An Encyclopedia of Myth and Legend.* London: Diamond, 1995.

Knutson, Roger M. *Furtive Fauna: A Field Guide to the Creatures Who Live on You.* New York: Penguin, 1992.

Lappé, Marc. *The Body's Edge: Our Cultural Obsession with Skin.* New York: Holt, 1996.

Laqueur, Thomas. *Making Sex: Body and Gender from the Greeks to Freud.* Cambridge, Mass.: Harvard U. Pr., 1990.

Lasky, Jesse L., Jr., and Frederic M. Frank, from treatment by Harold Lamb and Vladimir Jabotinsky. *Samson and Delilah* (filmscript). Hollywood: Paramount, 1949.

Latteier, Carolyn. *Breasts: The Women's Perspective on an American Obsession*. New York and London: Haworth, 1998.

Le Guerer, Annick. *Scent: The Mysterious and Essential Powers of Smell*. New York: Turtle Bay/Random House, 1992.

Leakey, Mary. *Disclosing the Past*. Garden City, N.Y.: Doubleday, 1984.

Levy, Mervyn. *The Moons of Paradise: Some Reflections on the Appearance of the Female Breast in Art*. New York: Citadel, 1965.

Lewin, Roger. *In the Age of Mankind: A Smithsonian Book of Human Evolution*. Washington, D.C.: Smithsonian, 1988.

Lockley, Martin. *The Eternal Trail: A Tracker Looks at Evolution*. Reading, Mass.: Perseus, 1999.

Lopate, Phillip. *Portrait of My Body*. New York: Doubleday, 1996.

Lord, M. G. *Forever Barbie: The Unauthorized Biography of a Real Doll*. New York: Morrow, 1994.

Lovejoy, C. Owen. "The Origin of Man." *Science*, 211(1981):341–48.

Lubell, Winifred Milius. *The Metamorphosis of Baubo: Myths of Woman's Sexual Energy*. Nashville: Vanderbilt U. Pr., 1994.

Lucie-Smith, Edward. *Sexuality in Western Art*. New York: Thames & Hudson, 1972, revised 1991.

Lutyens, Mary. *Millais and the Ruskins*. London: Murray, 1967.

Manguel, Alberto. *Reading Pictures: A History of Love and Hate*. New York: Random House, 2000.

Mazzio, Carla. "Sins of the Tongue," in *The Body in Parts: Fantasies of Corporeality in Early Modern Europe*, edited by Hillman et al., q.v.

McDonough, Yona Zeldis, editor. *The Barbie Chronicles: A Living Doll Turns Forty*. New York: Touchstone/Simon & Schuster, 1999.

McFarland, David, editor. *The Oxford Companion to Animal Behavior*. Oxford: Oxford U. Pr., 1982.

McManus, Chris. *Right Hand Left Hand: The Origins of Asymmetry in Brains, Bodies, Atoms and Cultures*. Cambridge, Mass.: Harvard U. Pr., 2002.

McNeill, Daniel. *The Face: A Natural History*. New York: Little, Brown, 1998.

Meyers, Jeffrey. *Orwell: Wintry Conscience of a Generation*. New York: Norton, 2000.

Miller, Geoffrey. *The Mating Mind: How Sexual Choice Shaped the Evolution of Human Nature*. New York: Doubleday, 2000.

Miller, Jonathan. *The Body in Question*. New York: Random House, 1978.

Miller, Russell. *Bunny: The Real Story of Playboy*. New York: Holt, Rinehart and Winston, 1985.

Mills, Jane. *Womanwords: A Dictionary of Words about Women*. New York: Free Press, 1992.

Milner, Richard. *The Encyclopedia of Evolution: Humanity's Search for Its Origins*. New York: Facts On File, 1991.

Montreynaud, Florence. *Love: A Century of Love and Passion*, translated by Simon Knight, Jean Pitt, and Mark Pallant Tripp. Paris: Editions du Chêne–Hachette Livre, 1997.

Morell, Virginia. *Ancestral Passions: The Leakey Family and the Quest for Humankind's Beginnings*. New York: Simon & Schuster, 1995.

Morgan, Elaine. *The Descent of the Child: Human Evolution from a New Perspective*. New York: Oxford U. Pr., 1995.

——. *The Descent of Woman*. New York: Stein and Day, 1972.

Morris, Desmond. *Intimate Behaviour*. New York: Bantam, 1971.

——. *The Naked Ape*. New York: McGraw-Hill, 1967.

Morrison, Toni. *The Bluest Eye*. New York: Holt, Rinehart and Winston, 1970.

Nabokov, Vladimir. *Speak, Memory: An Autobiography Revisited*. New York: Putnam, 1966.

Napier, John. *Hands*. New York: Pantheon, 1980.

Nekrasova, Ksenya. "The Blind Man," translated by Vera Rich, in *Twentieth Century Russian Poetry: Silver and Steel*, edited by Albert C. Todd and Max Hayward with Daniel Weissbort. New York: Doubleday, 1993.

Nelson, Geoffrey K. *Spiritualism and Society*. London: Routledge & Kegan Paul, 1969.

Nuland, Sherwin B. *Leonardo da Vinci*. New York: Viking, 2000.

Ostwald, Peter. *Schumann: The Inner Voices of a Musical Genius*. Chicago: Northeastern U. Pr., 1985.

Ovid. *Metamorphoses*, translated by A. D. Melville. Oxford: Oxford U. Pr., 1986.

Paglia, Camille. *Sexual Personae: Art and Decadence from Nefertiti to Emily Dickinson*. New Haven: Yale U. Pr., 1990.

Paley, Maggie. *The Book of the Penis*. New York: Grove, 1999.

Paley, William. *Natural Theology, or Evidences of the Existence and Attributes of the Deity Collected from the Appearances of Nature*. Oxford: Oxford U. Pr.: J. Vincent, 1828, second edition.

Palmer, A. Smythe. *The Samson-Saga and Its Place in Comparative Religion*. London: Isaac Pitman & Sons, 1913. Reprint, New York: Arno, 1977.

Partridge, Eric. *Origins: A Short Etymological Dictionary of Modern English*. New York: Macmillan, 1966.

Peck, Stephen Rogers. *Atlas of Facial Expression: An Account of Facial Expression for Artists, Actors, and Writers.* New York: Oxford U. Pr., 1987.

Peiss, Kathy. *Hope in a Jar: The Making of America's Beauty Culture.* New York: Metropolitan, 1998.

Petitto, Laura Ann, et al. "Language rhythms in baby hand movements." *Nature,* 413, 35–36 (2001).

Pinchot, Roy B., editor. *Muscles: The Magic of Motion.* New York: Torstar, 1985.

Pliny the Elder. *Natural History: A Selection,* translated and edited by John F. Healy. London: Penguin, 1991.

Podolsky, Doug M. *Skin: The Human Fabric.* New York: Torstar, 1984.

Posner, Gary P. "The Face Behind the 'Face' on Mars: A Skeptical Look at Richard C. Hoagland." *Skeptical Inquirer,* November/December 2000.

Proust, Marcel. *In Search of Lost Time,* six volumes, translated by C. K. Scott Moncrieff and Terence Kilmartin, revised by D. J. Enright. New York: Modern Library, 1992.

Pullar, Philippa. *Consuming Passions.* Boston: Little, Brown, 1970.

Purcell, Rosamond. *Special Cases: Natural Anomalies and Historical Monsters.* San Francisco: Chronicle, 1997.

Quammen, David. *Natural Acts: A Sidelong View of Science and Nature.* New York: Nick Lyons, 1985.

Ragas, Meg Cohen, and Karen Kozlowski. *Read My Lips: A Cultural History of Lipstick.* San Francisco: Chronicle, 1998.

Rhodes, Henry T. F. *Alphonse Bertillon: Father of Scientific Detection.* New York: Abelard-Schuman, 1956.

Ridley, Matt. *The Red Queen: Sex and the Evolution of Human Nature.* New York: Macmillan, 1993.

Robbins, Tom. *Even Cowgirls Get the Blues.* Boston: Houghton Mifflin, 1976.

Robinson, Julian. *Body Packaging: A Guide to Human Sexual Display.* Los Angeles: Elysium, 1988.

Rodgers, Joann Ellison. *Sex: A Natural History.* New York: Times, 2002.

Rose, Phyllis. *Parallel Lives: Five Victorian Marriages.* New York: Knopf, 1984.

Rutledge, Leigh W., and Richard Donley. *The Left-Hander's Guide to Life.* New York: Plume, 1992.

Sagan, Carl. *Broca's Brain: Reflections on the Romance of Science.* New York: Random House, 1979.

———. *The Demon-Haunted Universe: Science as a Candle in the Dark.* New York: Random House, 1995.

Sapolsky, Robert. *A Primate's Memoir: A Neuroscientist's Unconventional Life Among the Baboons*. New York: Scribner's, 2001.

Schama, Simon. *The Embarrassment of Riches: An Interpretation of Dutch Culture in the Golden Age*. New York: Knopf, 1987.

Sennett, Richard. *Flesh and Stone: The Body and the City in Western Civilization*. New York: Norton, 1994.

Sheets-Johnstone, Maxine. "Hominid bipedality and sexual selection theory." *Evolutionary Theory*, 9 (1), 57–70.

Siegel, J. M., et al. "Monotremes and the Evolution of REM Sleep," in *Philosophical Transactions of the Royal Society*, 353 (1998):1147–57.

Silverman, Kenneth. *Houdini!!! The Career of Ehrich Weiss*. New York: HarperCollins, 1996.

Simpson, George Eaton. *Religious Cults of the Caribbean: Trinidad, Jamaica, and Haiti*. Rio Piedras, Puerto Rico: Institute of Caribbean Studies, 1980.

Skal, David J. *Screams of Reason: Mad Science and Modern Culture*. New York: Norton, 1998.

Solnit, Rebecca. *Wanderlust: A History of Walking*. New York: Viking, 2000.

Spiegel, Maura, and Lithe Sebesta. *The Breast Book: An Intimate and Curious History*. New York: Workman, 2002.

Steinberg, Leo. *The Sexuality of Christ in Renaissance Art and in Modern Oblivion*. Chicago: U. of Chicago Pr., revised edition, 1983.

Strossen, Nadine. *Defending Pornography: Free Speech, Sex, and the Fight for Women's Rights*. New York: Scribner's, 1995.

Stücker, M., et al. "The cutaneous uptake of oxygen contributes significantly to the oxygen supply of human dermis and epidermis." *Journal of Physiology*, 538 (2002).

Tadié, Jean-Yves. *Marcel Proust: A Life*. New York: Viking, 2000.

Tattersall, Ian. *The Fossil Trail: How We Know What We Think We Know About Human Evolution*. Oxford: Oxford U. Pr., 1995.

———. *The Human Odyssey: Four Million Years of Evolution*. New York: Prentice-Hall, 1993.

Taylor, Timothy. *The Prehistory of Sex*. New York: Bantam, 1996.

Thompson, C. J. S. *The Hand of Destiny: The Folk-lore and Superstitions of Everyday Life*. London: Rider, 1932.

Thurer, Shari L. *The Myths of Motherhood: How Culture Reinvents the Good Mother*. Boston: Houghton Mifflin, 1994.

Torbrügg, Walter. *Prehistoric European Art*. New York: Abrams, 1968.

Tudge, Colin. *The Variety of Life: A Survey and a Celebration of All the Creatures That Have Ever Lived*. Oxford: Oxford U. Pr., 2000.

Twain, Mark. *The Tragedy of Pudd'nhead Wilson and the Comedy of Those Extraordinary Twins.* London and New York: Oxford U. Pr., 1996.

Unterman, Alan, editor. *Dictionary of Jewish Lore and Legend.* New York: Thames & Hudson, 1991.

Updike, John. "The Disposable Rocket," reprinted in *The Male Body: Features, Destinies, Exposures,* edited by Laurence Goldstein. Ann Arbor: U. of Michigan Pr., 1994.

———. "Get Thee Behind Me, Suntan," reprinted in *More Matter.* New York: Knopf, 1999.

———. *Toward the End of Time.* New York: Knopf, 1997.

Vaughan, Christopher. *How Life Begins: The Science of Life in the Womb.* New York: Times, 1996.

Vera, Yvonne. *Without a Name* and *Under the Tongue.* New York: Farrar, Straus and Giroux, 2002.

Vertosick, Frank T., Jr. *Why We Hurt: The Natural History of Pain.* New York: Harcourt, 2000.

Vickers, Nancy J. "Members Only: Marot's Anatomical Blazons," in *The Body in Parts: Fantasies of Corporeality in Early Modern Europe,* edited by Hillman et al., q.v.

Vidal, Gore. *Virgin Islands: Essays 1992–1997.* London: André Deutsch, 1997.

Vienne, Véronique, quoted in *Read My Lips,* by Meg Cohen Ragas and Karen Kozlowski, q.v.

Vogel, Steven. *Prime Mover: A Natural History of Muscle.* New York: Norton, 2001.

Vonnegut, Kurt. *God Bless You, Mr. Rosewater; or, Pearls Before Swine.* New York: Delacorte, 1965.

Wallace, Anne D. *Walking, Literature, and English Culture: The Origins and Uses of Peripatetic in the Nineteenth Century.* Oxford: Clarendon, 1993.

Ward, Geoffrey C. *Closest Companion: The Unknown Story of the Intimate Friendship Between Franklin Roosevelt and Margaret Suckley.* Boston: Houghton Mifflin, 1995.

Warner, Marina. *From the Beast to the Blonde: On Fairy Tales and Their Tellers.* New York: Farrar, Straus and Giroux, 1995.

Waters, Anita M. *Race, Class, and Political Symbols: Rastafari and Reggae in Jamaican Politics.* New Brunswick, N.J.: Transaction, 1985.

Watson, Lyall. *Jacobson's Organ and the Remarkable Nature of Smell.* New York: Norton, 2000.

Watterson, Bill. *The Days Are Just Packed.* Kansas City: Andrews McMeel, 1993.

Wheeler, Peter E. "The thermoregulatory advantages of hominid bipedal-ism in open equatorial environments: The contribution of increased convective heat loss and cutaneous evaporative cooling." *Journal of Human Evolution*, 21 (1991):107–15.

White, Edmund. *Marcel Proust*. New York: Viking, 1999.

White, Michael. *Leonardo: The First Scientist*. New York: St. Martin's, 2000.

Whitehouse, J. Howard. *Vindication of Ruskin*. London: Allen and Unwin, 1950.

Whitfield, John. "Goldfinger held grain of truth." *Nature* online, 13 February 2002.

Whitford, Frank. *Gustav Klimt*. London: Collins & Brown, 1993.

Willis, Delta. *The Hominid Gang: Behind the Scenes in the Search for Human Origins*. New York: Viking, 1989.

Willis, Roy, general editor. *World Mythology*. New York: Holt, 1993.

Wilson, Dudley. *Signs and Portents: Monstrous Births from the Middle Ages to the Enlightenment*. London: Routledge, 1993.

Winterson, Jeanette. "A Porn Reader." *Guardian*, 5 December 2000.

X, Malcolm, and Alex Haley. *The Autobiography of Malcolm X*. New York: Grove, 1964.

Yalom, Marilyn. *A History of the Breast*. New York: Knopf, 1997.

Zola, Émile. *The Debacle*, translated by and excerpted in Richard Gordon, *An Alarming History of Famous and Difficult Patients*, q.v.

INDEX

Jarvis, Robin, 304
Jesus, 108, 118, 158, 179, 249, 250
Jews, 37, 106, 249
Joblot, Louis, 49
John Paul II, pope, 116
Johnson, Celia, 57
Johnson, Samuel, 206, 244, 252, 276, 310
Johnson, Virginia E., 255
John the Baptist, Saint, 34
Jolly, Alison, 231–32
Jones, Chuck, 166
Jones, Frederic Wood, 292–93
Jones, James, 120
Jones, Mary Harris "Mother," 303
Joubert, Laurent, 202
Joyce, James, 59, 116–17, 213, 220
Judas, 118
Julius Caesar (film), 16–17
Jungle Book, The (film), 276
Jupiter and Thetis (Ingres), 21
Juvenal, 174

Kabuki, 66
Kahlo, Frida, 69
Kama Sutra, 254
Karma, 86
Kasson, John F., 140
Keats, John, 128
Kemal, Mustafa (Atatürk), 141
keratin, 27, 299
Khalil Bey, 262
Khoi ("Hottentots"), 284, 285, 286
Kikiyu, 38
King, Martin Luther, Jr., 303
Kiss, The (Rodin), 125–26
kissing, 116–28; film titles, 125
Klawans, Harold, 295–96
Kleitman, Nathaniel, 75–76
Klimt, Gustav, 125, 263, 264
Knight, Damon, 99
Knutson, Roger M., 260–61
Kohts, Nadjeta, 54
Komachi Hair Company, 259
Kozlowski, Karen, 112
Kruszelnicki, Karl, 216–17

labia, 107, 238, 241–42, 265, 285
lachrimal gland, 73
Laetoli footprints, 5, 269–70, 273, 310

Lagerfield, Karl, 60
Lake, Veronica, 26, 28
Lamarr, Hedy, 35, 219
Lamb, Caroline, 264
Landi, Neroccio de', 71
Landis, Carol, 219
Landseer, Edwin, 147–48, 154
lanugo, 23–24
La Pérouse, Jean-François de, 107
Lappé, Marc, 16
Laqueur, Thomas, 202
Larson, Edward, 308
Larson, Gary, 148, 276
Lascaux Cave, France, 247–48
Lavater, Johann Kaspar, 100
Lawrence, Jacob, 154
Lawson, Robert, 99
Leakey, Louis, 162
Leakey, Mary, 5, 270, 273
Le Brun, Charles, 68–69
left-handedness, 177–85
Left Hand, Right Hand! (Sitwell), 175
Leigh, Janet, 58
Le Mésangère, 98
lemurs, 163, 176
Lennon, John, 145
Leonardo da Vinci, 59, 60, 178, 202, 242, 246–47, 287–88
Leopold, Aldo, 42
leprosy, 161–62, 184
Levy, Mervyn, 210–11
Liam (film), 261
Liberty Leading the People (Delacroix), 206
lice, 260
Lichtenberg, Georg Christoph, 271
Lilli (cartoon character and doll), 203–5
lingam, 241, 248
Linnaeus, Carolus, 27, 193
lipreading, 132
lips, 103–28; and kissing, 116–28
lipstick, 112–14, 115
Little Dorrit (Dickens), 114–15
Little Richard (Richard Penniman), 41
Livingston, David, 108
lobster, 167
Lockley, Martin, 292
Logier, Johann Bernard, 159
Lolita (Nabokov), 110
Lombroso, Cesare, 69, 87–89, 180, 181
Longfellow, Henry Wadsworth, 41